Management for Professionals

For further volumes:
http://www.springer.com/series/10101

Taco C.R. van Someren
Shuhua van Someren-Wang

Green China

Sustainable Growth in East and West

Springer

Taco C.R. van Someren
Shuhua van Someren-Wang

Ynnovate
Hilversum
The Netherlands

ISSN 2192-8096 ISSN 2192-810X (electronic)
ISBN 978-3-642-28809-8 ISBN 978-3-642-28810-4 (eBook)
DOI 10.1007/978-3-642-28810-4
Springer Heidelberg New York Dordrecht London

Library of Congress Control Number: 2012950868

Printed on acid-free paper

Springer is part of Springer Science+Business Media (www.springer.com)

Foreword

The greening of industry has been on the agenda of corporate and government leaders in the West for many years. For some, sustainability is happening too slowly, while for others, it is not a major priority. Sustainability is a process—generally considered a long-term process—of global behavior change, particularly in our way of doing business. Over the last few years, we have taken the low-hanging fruit. Now it is time to reach higher.

Against the backdrop of global financial and economic crises and austerity, we cannot afford the luxury of thinking about sustainability as a long-term goal. There is also a strong short-term economic rationale for greater sustainability. The green technologies developed by Dutch companies in, for instance, waste and water management, renewable energy, and recycling have given them a leading position on these fast-growing global markets.

Sustainable enterprises generate profits and jobs. A recent EU report calculated that the full implementation of existing waste legislation by EU member states would create 40,000 jobs and save €72 billion a year compared with doing nothing.

For several years now, the Dutch Ministry of Infrastructure and the Environment has been investing in its relations with China. The two countries enjoy close, productive cooperation in both the public and private sectors. On my visits to China, I learned that the Chinese government is serious about greening its economy and eager to work on this challenge with international partners. This means there are many opportunities in China for Western companies to do business and ride the green wave.

Economic growth and sustainable development are two sides of the same coin. Sustainability can mean commercial success and support a society's long-term prosperity. This line of thinking is one of the pillars of the sustainability policy that I have successfully implemented in the Netherlands.

I like the way this book illustrates how business can work with the government in the interests of a more sustainable society. The authors show that both private companies and public organizations have to create and implement strategic innovations. I agree with their assertion that only those societies, industries, and organizations that are prepared to reinvent themselves along the lines sketched in this book will ultimately belong to the winning team.

We know that China is learning the lessons of our own experiences, good and bad, in our quest for a strong, greener economy. But increasingly, the Netherlands is also learning more from China. This two-way exchange demands mutual understanding. The authors highlight the fact that, despite numerous contacts and business deals between China and the West, there is still plenty of scope for improving relations and developing opportunities on both sides that benefit people, the planet, and profits.

I congratulate both Professor Taco van Someren and Dr. Shuhua van Someren-Wang on their efforts to integrate sustainability, strategic innovation, and China into a business perspective. This book again emphasizes that vital decisions need to be taken in the interests of the planet on which we live.

Joop Atsma
State Secretary for Infrastructure and the Environment

Contents

A New Green Growth Wave Led by China? 1

Sustainability and China at the Basis of a New Industrial Revolution

The green of China is not only for the sake of the planet but also for the long-term profit of the Chinese people. The European Union (EU) and the USA have no strategic answer to the greening of China. The world order is changing. The old growth curve is declining, and the old superpowers are being replaced. In the history, every time when this happened, strategic innovation was the fuel of change. Only those who could create a new way of doing business could go on. The indicators of a new world order are sustainability and the rise of China. China is trying to catch the new growth wave by turning into a sustainable economy. The driving forces of the New China have a deep-rooted cultural background which fit to the notion of sustainability. While the West is fighting the short-term economic crisis, China is rolling out this long-term economic plan to create sustainable enterprises for the future prosperity. The West is not prepared. Most companies still do not see sustainability as business while China is creating the biggest green business. The Western governments have neither a good green policy nor a good China policy. We are facing the danger of losing our welfare if we cannot catch up with the greening China. Luckily, the regained power of China is the result of a combination of several groups represented by different colors instead of only green. The rise of China is the start of a new industrial area in which political, social, ecological, economic, institutional, and cultural aspects of growth and development differ from the past. The USA, the EU, and China are on a crossroad. The West still has a chance if we can innovate our government and industry. Strategic innovations are the solution to cope with the new industrial revolution.

1.1 Shaking and Shaping

China is shaping a new future by turning green. China's plan for a sustainable future will have a big impact for the world economy. The West does not see the economic potential of either sustainability or China, let alone an answer to deal with it.

T.C.R. van Someren and S. van Someren-Wang, *Green China*,
Management for Professionals, DOI 10.1007/978-3-642-28810-4_1,

China's political and business leaders also have a clear idea about the role and place of China in the future world and its corresponding roadmap. Future China will have green as a major color. The aim of Green China is to become number one in technology and sustainability independent of the West. The plan will be put into action by pumping billions into the green business, tightening up the environmental regulation, and forcing mergers and acquisitions to create national champions. Other than normal business, green economy needs intervention of the government. And the West? The governments are mostly paper tigers with short- (election-) term targets rather than long-term economic strategy. In his State of the Union 2011, President Obama said that the USA needs a Sputnik moment. The President of the European Commission Barroso in his State of the Union 2011 said that significant growth is not an impossible dream. At the same time, many Western companies in the green sector have to fight for survival crisis after crisis.

China grows for decades, and the Chinese Sputnik has already been launched in 1978 by Deng. China showed the capability of a programmatic approach and a symbioses between public and private economy. This was the first step on the long road to restore China's historical world leading role and future world dominance. Time enough for the West to react and act. The only answer came from private industry which boiled down to outsourcing and Chinese market penetration. China is already earning money with the green technologies in Europe.

Despite cost-saving outsourcing tactics, the West has no adequate China strategy. In 2010, the EU presented its Strategy 2020 after its disastrous Lisbon Strategy 2010 with 10 years of bureaucratic sleep. And the USA? The USA gave prirority to twenty years of military war instead of economic fighting. In the most plans of both the US Administration and the EU Commission, China is somehow mentioned but mostly wrongly interpreted. No strategic answer has been formulated except old innovation mantras in a new jacket. In fact, nothing else than an improvement of old policies but no in-depth analysis of the consequences of the Chinese rise and its geopolitical, geo-economical, and geo-social consequences. In other words, the West produced words and paper but no sound grand strategy how to deal with the greatest change and shift in recent world history.

Does the failing Western strategic answer mean that the West is lost? No. Green China is not a divine creation without flaws, mistakes, and weaknesses. But it does offer innovative ways of doing business and needs new answers from the West. We are trying to point out the challenges and show the possibilities of a new path.

The Old World has to be shaked, and a New World will be shaped—by China, inside China, and outside China. Because of globalization of the world economy in the past decades and the Internet revolution, somehow, the idea has been widespread that we live in a village. Some call the world flat. Others call it one world.[1] This one world is similar to notions like harmonious society and peaceful development put forward and promoted by the central government of China. In this one

[1] See T. L. Friedman 2005, The World is Flat and E. Tse 2010, The China Strategy: Harnessing the Power of the World's Fastest-Growing Economy.

world, China will get a place, and we need an answer to that. Yes China will get a place, and we truly need a strategic answer.[2] But in our book, there is no one world, and it is definitely not flat. People like simplicity and simple solutions. But with the rise of a complex society like China, there will be no simple answers. Do not be afraid, we do not give complex answers, but we do need to go in depth. Otherwise, short-term thinking, crisis, and disappointment are the result. China and the West have different speeds, different performance criteria, different institutional systems, different norms and values, different behavior, and different views on the world. When we do not really understand China, huge challenges appear like mountains on the economic horizon, but if we can see where China and the world are going, new industries and new firms will emerge alongside the Himalaya.

Five mega developments determine the world's future and the fate of Chinese and Western societies:

1. A new world order emerges.
   The growth of emerging countries and China in particular reshuffles Western-dominated world order. On all levels of (trans)national governments, international institutions, industries, networks, and single firms, the domination of Western views and ways of working will be adapted, abandoned, and replaced by new ones.

2. New "market" rules are being created.
   Western market rules like free market and trade, fair competition, and shareholder value for single forms will be adapted and replaced by new rules focused on control and power relations. The Western distinction between macro level (political roles, economics, institutions), meso- (industry-level organization), and microeconomics (individual organizations) will be blurred. In fact, the Western industrial organization will be challenged and replaced or at least complemented by Asiatic or even Chinese models.

3. Sustainability is only one of the relevant issues.
   Greening of industry is a huge challenge for every current economy and even society as a whole. But Green China implies more than greening industry; it also comprises different market behavior, new priorities, and different business participants with different strategies and new behavior.

4. Technologies do not make the difference, non-technological innovations will.
   Of course new technologies will appear, but fast imitation strategies dwarf the relevance as a competitive advantage. Internet and social media do not cause revolutions; they are only instruments with new impacts. Intensity and scope of change is determined by nontechnical innovations such as drivers for behavior and ways of making decisions. Uniqueness can only be achieved

[2] See van Someren, StrategischeInnovationen, 2005a.

by focusing on nontechnical innovations. The challenges and opportunities of, for example, urbanization and megacities, different modes of securing resources, network versus total integrated firms, and institutionalized innovation will determine bottom-line results of societies and firms.

5. People.
   People are more than demographic statistics about population size and diversity. The historical background, size, age, diversity, and way of communication of people will profoundly influence world views, growth, development, organization and decision-making process of societies, relations between nations and regions, industries, networks, and firms. These human being-related aspects will substantially influence the nature of strategic innovations besides economic criteria. Cross-cultural behavior will be key.

The future world depends on the interpretation and appliance of these trends in strategic thinking and daily business.

1.2 Western and Chinese Ways of Development

Western and Chinese perspectives on development can be complete opposite as Fig 1.1 and Table 1.1 show.

Figure 1.1 gives some striking differences between our fundamental and often subconscious thinking, behaving, interpreting, and deciding. The roots of these different ways go back far in history and come back in our norms and values such as philosophy or religion, educational systems, bringing up of children, and history of economic development. Especially in the past centuries, the glorious time of the rise of western countries, the scientific approach of problems, and finding solutions dominated governmental and corporate leaders and decision making. This approach is characterized by analysis and dividing issues in small parts, the basic particles of our value creating system. Specialization, focus, division, and niches are the accordingly economic notions. Confrontation and thinking in good and bad, right and wrong, are the instruments of creative destruction. Our actions are linear in time and sequential.

The Chinese way often shows paradox characteristics. The specialist can turn into a holistic leader; business focus can be accomplished by gambling diversities. Mainstream market and niches can be combined. Patience is important, but as soon as it is time to act, they can act very fast. Trust is more important than truth. Harmony is better than confrontation, but not always. It depends on the situation. And this characteristic of the capability to adapt leads to flexibility. There is no one truth; it depends in the situation or time period. The future is evolving and developing in circles of time which makes history very relevant. The circular time notion of the Chinese makes them possible long-term strategist. It is easier for them to oversee the whole big picture in which harmony with environment and time flow is the most important. Harmony with nature is actually a very old Chinese principle. Only the last decades, the economic interests have

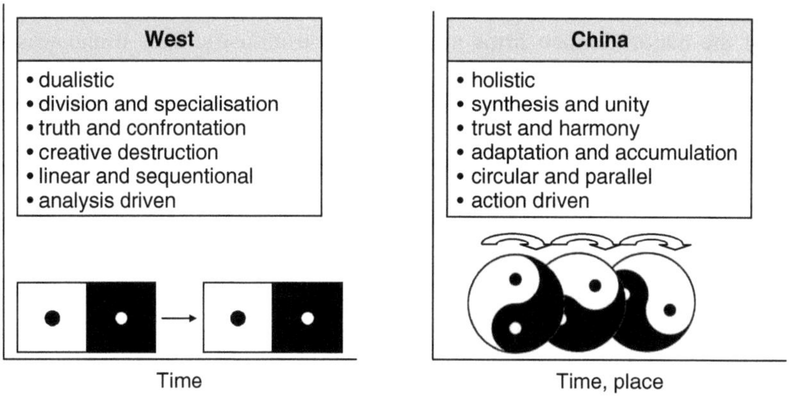

Fig. 1.1 Western and Chinese perspectives on development

Table 1.1 One global world: a question of perspective

How the West see Chinese	How the Chinese see the West
Be yourself, show our capabilities and the Chinese will like to do business with us	Western people cannot be trusted; they cheated us in the past and will do so in the future
China is an additional market which cannot be ignored because of the size of the market and the expected leading role of China	The West wants to exploit China and grab the profits
Transferring the newest knowledge to China is dangerous for our future position	The West has a large stock of new knowledge which is needed in China
China copies our technologies	The West with their technologies dominates the majority of our industry sectors
Western knowledge has to be protected against the imitating Chinese	China is a developing country; the rich, developed West should help China and transfer knowledge. This will help to create a harmonious world
Knowledge belongs to individual persons and institutions. Intellectual property rights are crucial for the innovation process	Knowledge is like fruit on a tree; it is available for everybody. Intellectual property rights are not important
Imitation of technology, products, and processes is a criminal deed	Imitation of Western technology, products, and processes is a sign of respect
Cooperation is in principle for benefit of all involved partners	Cooperation is to learn from the other and eventually to start for your own
The world will develop toward an open transparent society in which knowledge is shared for mutual benefit	China will regain its natural historical position being the center of the world and dominating the world order

overshadowed this principle. Now, the Chinese understand that they cannot go on developing their country in the old way. They learned this from the Western history. The Chinese always try to look for lessons from the history for the future. An old friend of ours working for a worldwide famous strategy consultancy bureau and responsible for the Asian market once got an assignment of the largest Chinese chemical companies to make a document of all historical strategic decisions of some of the major German firms in the chemical industry. This friend was very amazed, and in his Western opinion, it was a useless exercise not contributing to the strategic challenges of tomorrow. But for the Chinese Board, it is very important to know the history, the learning moments of success and mistakes, and to make acquaintance with Western corporate development in the long run.

When we apply these fundamental notions and current daily views on our global world, the different perspectives become clear. Table 1.1 gives a few illustrations.

From Table 1.1, we can learn that the Western view of China despite decades of doing business is still often distorted by parroting others or incorrect knowledge. But the other way round is also true. Western business people get cross-cultural training to try to adapt to Chinese circumstances. Most Chinese traveling in the West also learned a lot about Western history and culture but still behave like being in China. Sometimes, it is because of the pride of the regained power; sometimes, it is a nature reaction of holding on to what they know.

Because of fundamental different thinking, Chinese have different opinions about knowledge protection and what a peaceful world means. Doing business is for the Chinese like fighting war. When war is fought, winning is the most important. To win the business war, business tactics are allowed. The Chinese diplomats fight for less CO_2 reduction, more money, and free technologies for the developing countries in every climate conferences. At the same time, China builds up an own CO_2 reduction industry using technologies developed elsewhere.

Sustainable development according to the principle of harmony between people, planet, and profit is most of the time finding solutions for paradoxes. Chinese are very capable of dealing with paradoxes based on their fundamentally different way of thinking. When they make plans for scientific development, they choose to combine the paradoxes by earning money from the green industry. The Chinese capability of dealing with paradoxes is trained from childhood on by their parents, schools, and work. Table 1.2 shows a few examples.

The emphasis in good moral makes the Chinese agree with climate treaties; the pragmatism enables China to develop the green industry at their own timing. Generally speaking, the Chinese society is rather collaborative than individual. The group orientation dominates heroic individual behavior. But individual performance can also be much more important than group behavior. For example, when doing exams at school for getting access to best universities, the individual performance of the student is driven to the maximum by parents and family. The development of a green economy requires both collective actions and individual excellency. China showed their strength in combining these two by large collective investment in solar and wind energy and let the thousands of entrepreneurs to try their luck in green energy. When the investment in green energy threatens to be overheated, the government comes into force by merging small suppliers and increases the concentration ratio in the industry.

The well-known patience of Chinese and Asiatic people in general can be exchanged for decisiveness as soon as the situation requires this. Formulating plans for sustainable development and carrying them out is such a reaction.

Table 1.2 Paradoxes

Moralism	vs.	Pragmatism
Respect for age (elderly)	vs.	Young and restless
Circular time	vs.	Punctuality
Group orientation	vs.	Individual performance
Hierarchy	vs.	Informal channels
Decisiveness	vs.	Patience
Guanxi	vs.	Efficiency
Harmony and "face"	vs.	Rude behavior
Negotiation	vs.	Deal breaking
Masculinity	vs.	Zen

But continued investments in dirty industries only to gain a profit are happening at the same time. The Chinese entrepreneurs show strong masculine mentality by always wanting to be the number one. But when they fail to make money, they can use "Zen" to reflex and to learn positive lessons from their failure. Around 2010, many small companies in the green energy sector faced problem because of the broken loan chain when the banks raised the interest rate. Some even went bankrupt. But now the Chinese banks opened up the money tape; those who could survive the bad market situation and endure the crisis returned from the death. In this book, we want to show how these paradoxes characterize the future development of Green China.

1.3 China's New Long Cycle of History

We have to take into account the different perspectives on same notions like strategy and innovation. There is only one conclusion possible: There will be no one world, only a new world. But how new will it be? The rise of China is not a revolution but a renaissance of the Chinese economic position in the world till about sixteenth century. Some even think, that the last 300–400 years is only a short intermezzo of different Western countries. The dream of Chinese leaders is to revive the 5,000-year-old notion of a harmonious society and to create a new one for the future. The long cycle of history and circular time is now in favor of China again. This revolutionary roadmap toward a harmonious society has many dimensions which have to be realized. Moreover, as in any revolution, many dimensions are being renewed simultaneously over a longer time period. Western politicians have been navel-gazing, and private industry limits itself to outsourcing and market development.

The reaction of Chinese and Western business leaders on Deng's opening and support of central government was different. Chinese businessmen started new enterprises and revived old ones with and without direct governmental support. State companies were selected to become showcases in their industries, and private companies were set up to grasp new opportunities. Not only low-cost products and low-end added-value products are the domain of Chinese companies. Meanwhile, in future industries like renewable energy, electrical cars, and fast train sectors, Chinese firms belong to sector leaders. Western firms, including those which were already active in China before the opening, at least tried to realize two very old dreams. The first dream of Western businessmen is to sell their products and services to every single Chinese. The dream continues by the idea that taken together the Chinese represent a multibillion market. The second dream was to follow their old habit of reducing costs by using cheap labor and doing less in pollution control. Their second dream has been realized by many organizations by means of outsourcing, but it is coming to an end while the labor cost and environmental regulation cost are rising now in China. The money-earning dream is still a huge challenge for the green sector. But is it still an opportunity? That depends on our reaction to the development of the Chinese supply side, labor costs, quality, governmental market policies for market entrance and competition laws and regulations, and so on.

And what do scholars say? Do they deliver in-depth answers? Most books about China deal with the history, cultural differences, fast economic development, and probable future dominance of China. They give insight into China and its historical and recent development. Others speculate about the future winners, and completely contrary opinions exist. But most Western, or Western-raised, scholars neglect Asian and particularly Chinese influences. Fukuyama (1992) predicted the end of history with Western capitalism as the end phase of human economic–institutional development. Of course, due to the rise of China, recently scholars try to grasp the potential of Asia. Kishore Mahbubani (2008) pays attention to the rise of Asia and the blindness of Westerners but gives no clue what Westerners should do. Rifkin sees a third revolution in which EU and not the USA will lead due to the European values which would better fit to the required values belonging to sustainable development. Belonging to the winners solely on basis of norms and values fitting to sustainability is a very small basis for survival. Nouriel Roubini, also nicknamed "Dr Doom" due to his bearish pessimistic outlooks, gave a complete other future role to the EU. During the Davos 2011 meeting, he said that the EU has no chance and China, India, and Brazil have best chances. According to Roubini the, Western policy out of the crisis is "still lend, pray, and hope."[3] Also in Davos, Zhu Min, a Chinese special adviser of the IMF, sees a movement from a two-speed pattern to a three-speed arrangement. Newly developed countries and the USA are picking up and the rest of the developed countries are lagging behind.[4]

Who is right? We do not know. But more importantly, the race has not been decided yet because we are in the middle of the competition. In this book, we presume that everybody still has a chance to be the winner. Dr Doom underestimates the potential strength of the EU, and Rifkin has a myopic view on the European values which might diminish as a leading value for the world order.

[3] Daily Telegraph, 151210.

[4] Financial Times, 260111.

However, many of these and similar views are seen either through Western eyes or through Chinese eyes, and both have difficulty to understand each other's world. This can be easily solved by combining experienced authors with life experiences in both worlds as we tried to do. From a content view, most books give no answer on the question what the reaction of the West should be and how China should develop.

Let us see how Europe's business reacts. BusinessEurope represents 41 leading national business federations, and they issued a study on the rise of China and the necessary answers of the EU.[5] The EuropeBusiness study mentions several issues China should change in order to foster EU–China relations in the field of subsidies, technology transfers, WTO commitments, and trade rules. These kinds of answers mentioned in the study can be characterized as defensive and will not help the EU to be competitive in the future.[6]

The West did not see the real way China is growing and developing. The West was busy with itself and convinced that the rising of other world regions could be tackled without greatly adapting the pretended superior Western way of thinking and acting. In the past decades since Deng, the Western focus was on several hypes such as the threat of Japan, growth of Asian tiger states, the IT revolution, new economy, social media, innovation as the magic stick, oil wars, climate change, and financial crisis. Several of these topics appeared to be a hype such as the dominance of Japan, the New Economy followed by the Internet bubble, and partial innovation topics like creativity sector. Other crises in Mideast and Arabian regions linked to oil and financial crisis consumed large amounts of government spending.

China did not play a great role in all these hypes. The hype of trade missions of nation states, lower governments, and even small cities to China is mostly not strategically oriented and coordinated. China strategies are largely absent, and if one exists, it is characterized by little or no symbioses between the governmental China strategy and business China strategy. For the most Western government, politics and human rights are the main concern. For most Western companies, China was an excellent outsourcing region and a future sales market. Much less attention was paid to China as a competitor or even the possibility of China building up a sustainable economy. Only old strategies will not provide the right answer to our common future.

What about China? China is gaining world economic and political presence. The reform makes China a red capitalistic system with strong symbioses between public and private sectors. China developed itself to the workshop of the world and elevated about 250–300 million Chinese from poverty into middle-class citizens. The price of the necessary economic development in the past decades is environmental pollution and the risk of unacceptable social inequality. Furthermore, being the manufacturer of the world within one generation is a tremendous act of

---

[5] BusinessEurope, Rising to the Challenge, October 2011.

[6] During the presentation of the study in Brussels on Friday 14 October 2011, the first author of this book made this remark when questioning the President of BusinessEurope and the European Commissioner for Trade Karel de Gucht.

economic development but is more or less low-margin business. Here, the issue of production factors and value conversion comes in the picture.

In the recent history, China has gained a dominant position in three of four fundamental production factors: cheap labor (production factor labor), cheap and abundant capital (production factor capital), and fast and flexible decision making (production factor time). Knowledge is still lagging behind but improved. Natural resources (coal, oil, gas, renewables, materials, land, water, and air) are the one production factor which will dominate future battles for growth and development, and this is the production factor less developed in China but with high potential for dominant positions from a strategic perspective.

What are the alternative possible strategic answers of the West? It is useless to speculate about future winners or to make scenarios which will be obsolete within a year. Much more relevant is to define the cornerstones of a practical strategy of the future and to be flexible in its content and rollout. This will be the focus in this book. But we need to abandon the Western view and interpretation and try to use Chinese eyes and interpretation for the formulation of our strategic answers. One fundamental way of seeing through Chinese eyes is to accept a manifold world in which different options are available instead of a unilateral single truth world. This is like a rainbow.

1.4 China's Rainbow

Sustainability is closely linked up with all the production factors. Labor is linked up with demographics and social issues, natural resources with depletion rate and climate, capital with sustainability criteria, and time with accumulation of environmental detrimental effects, knowledge, and timing of decisions. Therefore, solving environmental problems and turning these problems into opportunities and creating high-margin industries will be the future basis of economic development thereby creating stronger social structures. It can be concluded that China also needs strategic innovations to keep the pace of GDP growth and extending prosperity to the rest of their population.

China's rainbow is a metaphor for the factors which can make China green if the Chinese use them wisely. The traditional basic colors are blue, red, and yellow. Green is mixture of the two basic colors blue and yellow. But the color red can change the green in a different shade. Each of these colors represents a certain aspect of society contributing to Green China. Table 1.3. gives an overview.

Chinese green will not be like Western green. The West and China have different histories and will make different decisions for the future. A rainbow range of colors will be the more possible mixture for China than only green. Rainbow China represents the mix of political, economical, institutional, social, and ecological developments of China's future. Blue China represents the available nonhuman resources like natural resources, available knowledge, and capital. Yellow China represents the increasing important role of civil servants, entrepreneurs, knowledge workers, and Chinese labor force. In times of the Chinese dynasties, imperial yellow was only allowed to be used by the Chinese emperor. Imperial yellow has

Table 1.3 Green China as a mixture of China's rainbow

China's rainbow	Content
Blue	Natural resources, capital and stock of knowledge, emissions, and pollution
Yellow	Civil servants, entrepreneurs, knowledge workers, and Chinese labor force
Red	Laws, regulations, policies, and norms and values
Green	Varying mixtures of blue and yellow and red

now been taken over by the Communist Party, and now they formulate policies, laws, and regulations. Together the colors yellow and blue mingle to green. However, this green is not completely finished. The reason is the color red. Red China represents the past developments of the Chinese social market economy in which governments take the lead by formulating the laws, regulations, boundaries of China's future system, norms and values, social system, networks, and innovation system. As a result, depending on the degree of influence from red China and market economy, the colors can be mingled in infinite combinations, and their number of color shades and color shifts are endless. How Green China will become? The future is not decided yet.

1.4.1 Red China

Red China needs a comprehensive law and order system. China is for 1,000 years of hierarchical society organized according to Confucianism. Chairman Mao has turned the communistic system very smartly into a Chinese hierarchy. Despite the opening up of China by Deng, the Communist Party kept their control over the country and their power as lawmaker and enforcer. But they adapted to changing circumstances by the programmatic reform. The deep red of communism order is gradually being replaced by a Chinese socialistic market system. It is a mixture of top-down policy making and creating a Chinese society and bottom-up initiatives within the boundaries of the Communist Party. In this respect, Red China is still a top-down governed state. For example, not only green policies and laws try to cope with sustainability issues within the country, but also forced mergers between windmill manufacturers are simply top-down ordered. For the future, this top-down government is also planned for the development of a worldwide leading green industry. This Red China can be maintained as long as hardware is the most important for business. But solving the sustainability issues will require a development in the soft part of the business too.

1.4.2 Yellow China

Yellow China is formed by its people and their collective social society. The people of China will discover that they need to have a sustainable way of development. Once they put their heart into it, the Chinese can achieve what they want by joining forces and driving themselves to the limit.

Among students from different nationalities in world famous universities, Chinese talents are well known. It is not because that Chinese are more intelligent. Most of the time, it is because they are ambitious, determined, and hardworking. Do one's best and learn as much as one can is a thousand-year-old Chinese mentality.

The development of the civil servants' awareness on sustainability is actually the drive behind the Chinese greening development policy. In the Chinese society, not all occupations are equal. For thousands of years, the most respectful occupation was civil servant. Established in 605, the Chinese imperial examination was the first examination system in the world designed to select the best administrative officials for the state's bureaucracy. This system created a class of scholar-bureaucrats irrespective of their family background. That is why Chinese parents put so much emphasis in good education of their children. This was the only way to rise in the ladder of Chinese society.

The most brilliant Chinese now still wanted to be a civil servant by taking strict entrance exams. Ten thousand candidates applying for one civil servant vacancy is very common in rich regions. Because of their different study backgrounds, a good mix of talents can be created. Because of the high quality of the civil servants, the administration system can be very efficient. On the other hand, being a businessman was always discouraged in the imperial China. Luckily, the opening up policy brings welfare to China. It also brings respect for being rich. For the first time in history, business activities are respected. This development provides the chance for those who do not fit in the mainstream. To get a good job, Chinese had to go to the good schools and start establishing a good network. The most important is to learn to fit in. Reading the stories of the first-generation Chinese entrepreneurs, you will find out that most of them were rebels or chanceless for regular jobs. That is maybe why they are more creative in business than normal Chinese.

Normal Chinese are hardworking people. If they are scholars, they have to work hard to be the best one to get funding for their research. If they are workers, they have to work hard to be the employee of the month for the extra bonus. It does not matter what they do, they always have to do their best. There is no sympathy for those who do not try their best. The only safety net is the family and friends. But the family and friends are also expecting that one day, they can lift with the success of a member of the family. A famous Chinese phrase is: "When a man achieves the Dao, his family and pets rise to heaven too."

The collectiveness enforced the individual ambition. The attitude of taking care for family, elderly, and network is a bridge to another characteristic of Yellow China. Many Chinese now have a dream to become founder of their own company and to be their own boss. When they cannot get support from the government or the bank, they can always get support from their network. The informal banking system is one of the drivers behind the overheated investment in, for example, LED light industry. But when the Chinese network is used right, it will enforce a trend toward commercialization and creating a Chinese market system. At the same time, the powerful public sector can be either corrupt by the network behavior or motivated by the collectiveness to support green business. The Chinese leaders are very aware of this. Therefore, in the future, anticorruption and the private and public symbiosis will be maintained and developed in new directions.

Yellow China also represents the trend toward increasing urbanization. The urban economy offers opportunities for new services and mobility systems. Important opportunities are the development of a health care system, a social security system, and pension plan system. The rising labor costs in China, the competitive pressures in low-margin industries, and the growing elite of successful entrepreneurs and rich government members are the cause of a growing gap between the rich and the poor. This growing social inequality represents a danger for Red China because it has the potential of eroding the legitimacy of the Red China pillars.

1.4.3 Blue China

The blue of China is unfortunately dirty blue. China became the factory of the world with the goodness and badness of manufacturing activities. China has too little environmental resources and capacity to continue to develop the old factory model of growth. Enforced Chinese influence in the world is sometimes driven by economic reasons like nature resources security rather than political reasons. China will regain its dominant position within the Asian region, and it will conquer new geographical areas by peaceful means to ensure the supply of necessary resources. This is Deep Blue China. But not all resources can be imported.

China has also paid a price. It is estimated that the unlimited annual economic growth of GDP resulted in a loss of 2–3 % of GDP due to ecological detrimental effects. Examples are polluted air resulting in health problems in urbanized regions and polluted rivers. One of the megatrends worldwide but also in China is the insufficient availability of fresh water. Table 1.4 shows the expected shortage of water in China, Brazil, India, and South Africa in 2030. Four of the five BRIC countries are involved, and in Russia, the water management is not much better.

Besides water, food is another crucial necessary condition for a prosperous development of society. Here, China has a similar scarcity problem as with water. The demand for food is greater than the supply of food. China will have to find a sustainable way out of their hunger to resources.

The rise of China and its focus on economic growth has resulted in several successes. For almost three decades GDP rose between 8 % and 12 %. About 250 to 300 million Chinese were levied into the middle class. Based on hard work, the accumulation of capital reserves resulted in capital reserve of more than 3.200 billion dollars. According to Naisbitt and Naisbitt (2010), this is the result of long-term continuous reform and political leadership like a China Inc. This interpretation is not true. Long-term plans and continuous leadership were indeed present, but they are only part of the whole picture. Moreover, with their opinion, the Naisbitts focus the success on Yellow and Red China. They forget that private corporations as part of Blue China are increasingly becoming important and that they contributed substantially to the economic success. Furthermore, the focus on long-term plans gives the impression that it is a single plan being implemented that lead to success. This is also not true. Part of Blue China is cutthroat competition between regions

Table 1.4 Supply and demand raw water

Country	Current supply (in billions of meters cubed)	Demand in 2030 (in billions of meters cubed)
China	618	818
Brazil	18.7	20.2
India	740	1,500
South Africa	15	17.7

Source: McKinsey, 2009

and individual organizations. Blue China also represents a very dynamic and big rivalry between organizations and networks within the society. This is a force independent of any central plan and following their own dynamic rules. Despite globalization, the local economy will become more relevant than in the last century.

1.4.4 Green China

There is no One China, and there is also no one Green China. Green China can have different shades depending on the mix of the colors. It represents the megatrends and the developments in society on all levels. For example, the past decades the mantra of Deng dominated behavior of market participants. Deng's well-known saying that it does not matter whether the cat is black or white as long as it catches a mice represents the unlimited economic expansion. In the future, some counter forces in China will emerge. The wealth creation at cost of environment and political legitimacy due to rising social inequality represents the core Triple P dilemma for the future of China. This is why Green China is not only a topic to be encountered from the ecological side but also requires strategic innovations. China needs strategic innovation on all levels, in both public and private sector, in every industry, and in every organization to find their new way of creating a sustainable future.

The West needs strategic innovation to find an answer to the raise of Green China. The governments need to create a good long-term China strategy with symbiosis with the private sector. The private sector needs to create ways to earn money from and with the raise of Green China.

1.5 Is Strategic Innovation Answer to Green China?

Understanding Green China implies understanding the macro, meso, and micro level of our world society. When we analyze from macro to meso to micro level, the following broad picture emerges.

From the long-term perspective of historians, the next decades will be dominated by the issue of sustainability as Fig. 1.2 illustrates. This is the macro perspective.

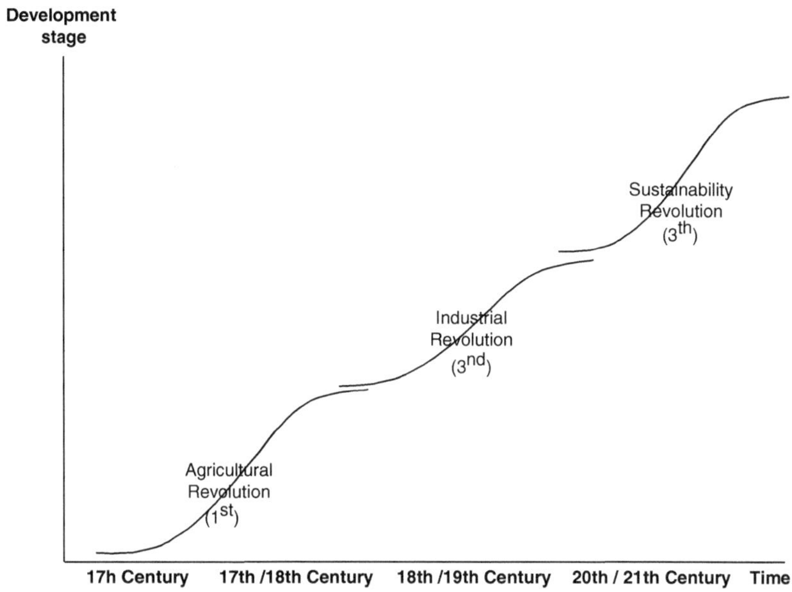

Fig. 1.2 Successive revolutionary waves

The wave or growth curve of sustainability is the successor of the agricultural and industrial revolution.

From an economic long-wave perspective, we can zoom in and see what happens in these historical long waves. Table 1.5 shows already some more detail.

On a macro level, the Kondratiev cycle of upswing and downswing spans a lifetime of 70–90 years. The new Kondratiev cycle started around 1990s represents the core issue of sustainability, the rise of China, and the slowdown or even downfall of the Western countries.

On the meso level, it is connected with, for example, radical innovations, rise of new industries, and new dominant countries. On the micro level, it is connected to the rise of new technical and nontechnical innovations, new firms, and new governance structures. The previous two long cycles are characterized by the domination of typical energy sources of coal and oil from leading dominant countries the UK and the USA, respectively. Therefore, it remains to be seen, for example, which dominant countries and energy sources will dominate the next decades. Will it be China and will it be wind, solar, biomass, tidal wave, hydro, hydrogen based on any source or nuclear (shale) gas, coal, and oil? Moreover, new dominant industries and key sector emerge as well as new dominant firms. Will it be Chinese firms such as Yingli or some corporations from other BRIC countries?

When we zoom in even further, the narrow interpretation of sustainability is about climate change, CO_2 and other emissions, resource depletion, renewable energy, and biodiversity. In Western societies, this is known as the greening of industry or society. In this perspective, sustainability boils down to dealing with

Table 1.5 Historical overview innovation paradigms

Mega trends	▪ mechanisiation ▪ factory organisation	▪ Inflexible water energy ▪ new transport systems	▪ standardisation ▪ Inflexible steam pwer ▪ 'unit'-Production	▪ flow production ▪ large scale production	▪ flexibility ▪ diseconomies of scale ▪ renewing value chain
Organisation	▪ Individual entrepreneur ▪ small Ffrms ▪ partners ▪ cooperation ▪ "Verlagsystem"	▪ small and medium sized firms ▪ legal obligations ▪ „Joint-stock" companies	▪ cartells, trust ▪ state owned (natural monopolies)	▪ multinational firms ▪ vertical integration ▪ divisionalform ▪ globalisation	▪ network ▪ alliances ▪ individual entrepreneur ▪ local scale
Industry	▪ textile ▪ textile chemicals	▪ steam engine ▪ steam railways ▪ railway suplliers	▪ electrical engineering ▪ maschine construction ▪ telegraph & telephone ▪ basic chemicals	▪ automotive ▪ consumer goods ▪ petrochemical ▪ oil	▪ ICT ▪ environmental technology
Material Energy	▪ water power	▪ coal ▪ steam power	▪ coal ▪ electricity	▪ oil and other fossil energy ▪ oil based chemicals	▪ electrical engine ▪ hydrogen engine ▪ renewables
Key factor	▪ cotton ▪ iron ore	▪ coal ▪ transport	▪ steel	▪ energy ▪ car	▪ chips ▪ bio technology ▪ battery technology

1770/80 **1838/43** **1880/90** **1930/40** **1980/90** **2030** **Time**

Source: Freeman, 1987; van Someren, 1991, 2005

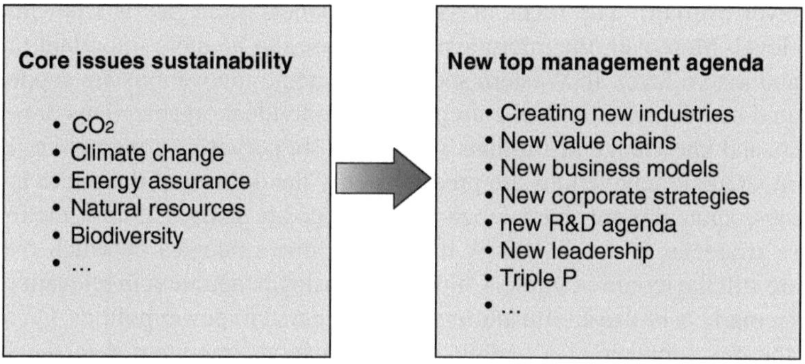

Fig. 1.3 Sustainability puts strategic innovations on top management agenda. (Source: Ynnovate 2003)

Triple P consisting of political, fiscal, institutional, economic, social, and ecological aspects of development. This is what most advocates of sustainable development in different variations tell us. However, this is only true when sustainability is seen as something we should somehow try to integrate in our business activities and market behavior. This is represented in Fig. 1.3.

Figure 1.3 shows that the main topics of sustainability indeed have a great influence on the management agenda of individual corporations, networks, and policy makers. Western corporate strategies are on this micro level. In this book, we will show that the other levels are becoming increasingly relevant due to the rise of China.

In the next decades, it is expected that China surpasses the USA as a leading economic nation. The focus of macroeconomics is on international trade relations, development of nations measured in, for example, GDP and currency issues, and economic policy for governments and international organizations like IMF. The main macro indicators in the West show a slow GDP growth whereas China will take the lead in total GDP. In the EU, we will see an increasing average age in demographic development and population reduction and in the USA, a growing population partly due to immigration. The accumulated huge debt in the USA and several European member states will slow economic growth. These macroeconomic indicators are more a result of activities—or absence of activities—than cause of slowing profits and decreasing wealth. No monetary measure can countervail these macro outcomes; only activities on the meso and micro level can keep the West in the race for wealth and prosperity. Meso economics is linked with issues like mergers, cartels, industry analysis, and policies. Business strategy is linked up with the micro level representing the strategies of individual firm. Business strategies are mainly formulated on a micro level (from the perspective of a singular organization trying to survive in a changing world). Innovation is in first instance also formulated on an individual organization level, but also, cooperation between organizations, that is, meso level is used to develop innovations. Sometimes, individual entrepreneurs or firms are able to create a new industry (micro- and meso-level overlap). The focus of Western business strategies is from micro to macro level. Moreover, the micro level is supposed to be more important than the meso and macro level. In Western societies, strategic innovations are supposed to be created mainly by individual entrepreneurs, individual organizations or network of firms, and knowledge institutions sometimes supported by government. Hence, Western strategic innovations are predominantly the domain of the micro level.

Chinese strategies are characterized by the opposite order. Till now, the focus of Chinese strategies is dominated by macroeconomic strategies of which five-year plans are still the iconic example. Chinese increasing penetration in relevant natural resource markets is also an illustration of Chinese macro power politics. On a meso level, the development of a certain industry is more important than the fate of individual firms belonging to these industries. On the contrary, individual regions and firms are selected with the aim to grow and strengthen key industries and China as a society. On micro level, classical strategic development in the Western sense is in China more or less in its infancy. For example, Chinese takeovers and new market penetration and developments are not as successful yet as their macro policies. For example, the entrance of the automotive sector by Landwind in Europe failed due to misjudgment of the role and relevance of car safety requirements in Europe. But Chinese business strategies of an individual firm are much more linked up with the meso and macro level than in the West. For example, in the renewable energy market, Chinese government forced the most important windmill manufacturers to merge in order to reduce overcapacity and to create world class players needed for realizing macroeconomic goals of going West macro policy.

For both the West and China, the greening of industry requires the involvement of all three levels, and this is the reason that the West and China can learn from each other. Because of the nature of sustainability, all three levels are relevant. Therefore, future strategies will have a completely different content and scope than existing traditional business strategies. The strategic innovation theory (SIT) presented later on in this book blurs the distinction between the micro, meso, and macro levels.[7]

Western style approaches can be divided into the American way (including the UK) or the European (continental European) way. The American way focuses on individual entrepreneurship and shareholder value. The European way is a more dirigiste governance of innovation and stakeholder approach (Rhineland model). However, differences in the interpretation of the Rhineland model between European countries exist. For example, France has a more centralized and top-down governance than the German innovation system. The financial crisis was the trigger to revive this discussion again because the shareholder approach connected with its short-term orientation and bonus system was regarded to be the central cause of the crisis. The huge support financial aid programs by member states to counterattack the crisis was another appearance of the Rhineland supporters.

[7] van Someren, 1991b, 2005a.

Additionally, discussions about the macroeconomic financial system, the depreciation of the Chinese currency renminbi or RMB, the survival of the dollar as worldwide leading currency, and the viability of the Euro cover the real foundation of prosperity. The rise of the renminbi (RMB) is not initiated by any world finance institution like Chinese Central bank, World Bank, or IMF, but its appreciation started due to an increasing entrepreneurial economy creating value. Of course, after that, Chinese government started to guide its national currency in the direction of an international currency. As of 2009, several measures were taken such as an experiment for cross border transactions; the monetary authorities of Hong Kong releases guidelines for trading of RMB, allowance for banks and later a limited number of firms to carry out transactions in RMB, and the allowance for Chinese firms to use RMB for mergers and acquisitions and other investments. It took China more than 30 years to build up a strong economy with high future earning potential. Instead of giving in on the demands of the USA to depreciate the RMB, the government of P.R. China issued on early March 2011 the policy to set first steps to let the RMB become the world's leading currency.

The core message of this book is that the real foundation of growth and prosperity is the ability to convert production factors into value by means of introducing strategic innovations (van Someren 1991, 2005). This is the value conversion principle. Only the factors like sustainability and rise of China are becoming the trigger factors for the next growth wave. Chinese do not care more about the environment by turning green. The biggest drive is to sustain the growth, or simply stated: Work hard, do something new, and earn money.

The Chinese foresee that the next new export market will be in the green sector. And as a sound bite, the long-term strategy is building up the earning capacity in the green sector. The Chinese are able to conquer world markets because they earned money by working hard toward their long-term strategy. In the case of Green China, earning money is easier because the Chinese way suits the green sector. The Chinese way is the symbiosis between firms and government with the aim to achieve nationwide long-term goals and at the same time short-term private business profit. The green sector is one which needs large-scale governmental involvement and long-term strategy. The short-term benefit will attract the private business.

What is the relevance for Green China? Let us look at the example of solar panels. The Chinese government created a market by announcing new energy policy in the five-year plans. The enormous demand for energy triggered Chinese entrepreneurs to start up solar energy firms most of the time with governmental support. Most of them first created local home markets within their provinces of origin with the local government as launching customer. Some of these firms grew rapidly by exporting to the subsidy-fueled European market. More entrepreneurs came in the renewable energy sector with private finance, and a new industry was born. Because of the wild growth, some of them including the first movers will not survive. Central government reacted by targeting some of these firms and industries as key technology and key sectors. The ones with less chance to survive will be forced to merge with the strong ones. Regional clusters of renewable firms have been created this way.

Since the financial and debt crisis in the USA and EU, central government has to rethink where to invest their received money. The traditional safe haven of state bonds has largely disappeared. So what are the alternatives? In this book, we will see that strategically innovating our way of doing business can provide a way to ride with the Green China wave. But both China and the West have a long way to go.

1.6 Green History of China

Looking at the green history of China, we will understand why China needs to be and will be sustainable. China's history has shown many different developments but also some leading principles. One of the leading principles is the notion of harmonious society. The idea of harmonious society is 5,000 years old. Harmony enters all levels of society. On a macro scale, the development of nation should be harmonious. On a meso level, the development of a network and the interests of network members should be harmonious. On a micro scale, everyone's life should be harmonious. Important to know is that harmony in Chinese context is not harmony at every single moment in life but should be interpreted as harmony in time. The former is the static view, the latter the dynamic view. The dynamic view fits to the notion of an emulation process. The duration of the time period depends on each situation. For Westerners, this is a very fuzzy and not crystal clear beacon for organizing one's life. For Chinese, it is. For example, in case of recent Chinese societal development, by purpose, the focus has been on economic development. The reasoning is that first economic development has to be achieved which is a necessary condition for solving any accompanying social and environmental problems. The unsustainable China is only one side of the medal or in a Chinese saying: First, fill the rice bowl even when it is dirty (being the environmental–social consequences of 30 years of economic reform). When most mouths are fed, other issues will gain importance. In time, harmony between these three items will be achieved.

The upcoming Chinese welfare and growing prosperity for growing parts of the population lead to another reality, the reality of pollution of water, air, soil, and increasing demand of energy and (scarce) resources. Contrary to what most Westerners think, since the 1970s, there is growing awareness in China about the detrimental effects of an unsustainable development and growth. The awareness appears in several ways.

One of the ways is research and education. Chinese scholars studied the environmental lessons the West learned. Books over the consequences of environmental disasters were published. Study on environmental sciences became a major in chemistry, geology, or biology departments in a few leading universities. At that time, very exclusive studies are only for the lucky few with good family connections or who are really excellent. The first-generation environmental activists were educated at that time. Many of them are now leaders in Chinese governmental organization. From 1990s on, international society put more and more pressure on China to improve its environmental policy and legislation. Environment issues became diplomatic issues.

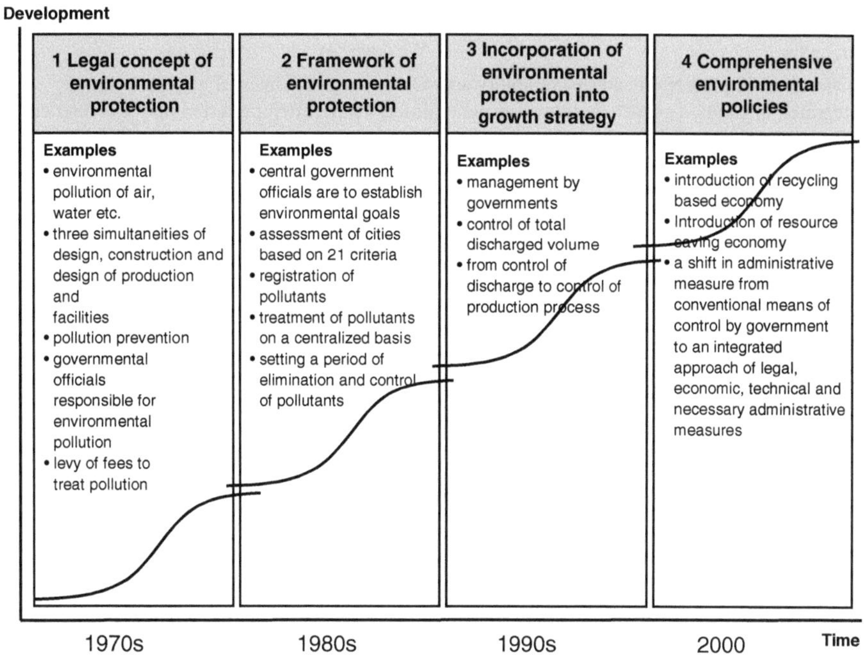

Fig. 1.4 Four stages of environmental policy shifts. (Source: stages based on H. Matsuno, 2009, p. 2-3)

Chinese environmental activists turned the pressure into their own force. Defending its way of development in every international encounter, they increased their status within the Chinese administration. At the same time, they use facts and figures to persuade the authority toward a more sustainable way. Economic cost of the damage to the environment is calculated. It has been accounted that the braking effect of unsustainable development reduces GDP growth with approximately 2–3 % per year. The fast development and growth lead to increasing income inequality and differences between rich and poor regions in China. The acquisition of resources for keeping up the fire of growth requires a power play on a global scale. International cooperation was and still is necessary to introduce new sustainable technologies and regulations. As in most countries, environmental laws and regulations did exist in China, but it has been renewed in the past decades as Fig. 1.4 demonstrates.

Matsuno (2009) differentiates between four stages starting in 1970s. Comparable to European countries and the USA, in the first stage, the focus was on environmental compartments like air and water. The second stage was characterized by formulating a framework of environmental protection. The big change with regard to addressing environmental issues came during the 1990s. In this third stage of policy development, environmental protection became part of Chinese growth strategy. The historical notion of harmonious development was introduced to indicate the Chinese way of sustainable development. In the fourth

Table 1.6 Strengths and weaknesses of Chinese environmental law and policies

Strengths	Weaknesses
Environmental protection and sustainability are integrated in China's growth and development policy and strategy	On lower governmental levels, economic interests are often preferred above ecological interests
Responsibility for sustainability is on highest hierarchical administrative and political level	Fast growth and development of economy cause backlog in environmental protection implementation and inadequate legislation
Sustainable policies and laws and regulations are one of the main drivers of innovation	Execution is weakest link
Political leader and high-ranked officials are personally made responsible for environmental protection results	Lack of management resources and experience
	Unclear coordination between administrations leads to inefficiencies and ineffectiveness
	Command control need to be complemented by incentive schemes

stage, the aim is to have a more comprehensive approach of environmental matters throughout all levels of Chinese society. The leading notions of "recycling-based economy" and "resource-saving society" were introduced in, respectively, 2004 and 2005. In both the 11th and 12th five-year plans of the central government of China, environment and sustainability issues and other related notions can be found back.

These developments in time show the rapid development of Chinese policies and legislation. China copied and adapted many laws and regulations from several European countries such as The Netherlands, Germany, and Sweden. But this is not a guarantee for achieving the aims written on paper. In Table 1.6, a few strengths and weaknesses of the Chinese policy and legislative system are mentioned.

Besides the content of the policies, the enforcement and execution are relevant. The adherence and enforcement are the weakest links with regard to environmental policy (Wang 2007). According to Wang (2007), some examples are legislative objectives remain unachieved, enforcement is superficial, excessive time exists between noncompliance and enforcement, available punishment for noncompliance is inadequate, injured parties are not properly compensated, and some environmental crimes receive administrative instead of criminal punishments. Besides these arguments linked with legislation itself, some other issues are relevant.

The relevance and impact of laws and regulations is also dependent on the political power within the governmental system. In this respect, the rapid increase of power and influence can be derived from the successive elevation of the department responsible for environmental policies. This development is more or less parallel to the stages mentioned in Fig. 1.3. There are four milestones of this institutional development (Xin and Honglin 2009). In the 1980s, triggered by the market system reform, the Environmental Protection Leadership Commission was succeeded by the Environmental Protection Agency (EPA) which took care of environmental policies. The Chinese EPA was a part of Ministry of Urban Construction and Environmental Protection indicating that protecting the environment was a relevant issue but had not first priority. Within a few years, this opinion changed, and in 1984, the EPA became an executive responsibility of the State Council and was renamed as National Environmental Protection Agency (NEPA). NEPA was nationwide responsible for management and supervision of environmental protection. The next step in the political-administrative hierarchy came in 1998 when the NEPA was directly organized under the State council. But this new State Environmental Protection Administration (SEPA) was not member of the cabinet and lacked voting rights in the State Council. This changed in 2008 when the SEPA attained the highest level and became the Ministry of Environmental Protection (MEP) being part of the cabinet and voting rights in the State Council.

When summing up the developments in the environmental policy arena, it can be concluded that in an early stage, Chinese leaders discovered and acknowledged the relevance of environmental protection and sustainability. But execution is a different matter. Moreover, priority of environmental issues against the interest of economic growth is also another matter. It demonstrates that the political system is divided in different groups with different aims and agendas and a Chinese way of consensus has to be found. The rainbow changes and determines the ultimate outcome. The end results of these power fights can be traced back in, for example, the five-year plans. All these issues will come back later in the coming chapters.

1.7 The Future Welfare: $€ or $¥€?

In the coming decade, it will be decided in which direction this new world order will be developed. At the moment, alternative directions of the path to the future are still possible. The scale, scope, and pace of this change should worry every other world player. When adequate reaction and action is taking too long, our welfare and well-being can be threatened. In the emerging new world, everyone has to conquer its place again. However, every country, region, and organization will generate its own strategic innovation fitting to its history, experiences, context, and ambitions. This holds for both the West as well as China. The reason is that in the past, our society has constantly been renewed, and if you believe in theory of stages, sustainability could be the fourth Industrial Revolution (agriculture, transport, manufacturing). But then, sustainability needs a broad front of innovations together called strategic innovations.

These stages mentioned in Table 1.5 are featured by the entrance of, for example, new materials, new sources of energy, new products and services, new organizational forms, and new firms. However, new technology dominates in this view. This stage model is a very common representation and description of events in these times of change. But in reality, innovation is a continuous process in which non-technological innovation is equally relevant as technology (van Someren 1991, 2005). Therefore, the stages are only a caricature of the reality. Much closer to the reality are small steps and accumulation of small steps leading to bigger steps. But also setbacks belong to the process, for example, by imitating and improving Chinese car manufactures learned from their Western business partners. As a next step, entering Western markets was a logical move. However, these first steps in case of Landwind were not a success due to neglecting safety issues.

It is possible to formulate an approach of innovation management which is independent of these stages and comprises technical and nontechnical innovations taken together called strategic innovations. The whole history of societal development and its accompanying development of the (market) economy is a process of emulation. Emulation comprises the will of individuals, organizations, and nations to be the best and surpass the other by means of any kind of small or big technical and nontechnical innovations. Innovative entrepreneurs are the driving force of innovation. Innovative entrepreneurs are able to make unique new combinations of production factors leading to new products, services, organizational forms, and business models. Innovation is about change and new things, but the history of innovation also offers a constant which is the common determinant of renewal. The determinants of any innovation are economies of scale, economies of scope, and economies of time. This constant or modus operandi of strategic innovation is called Triple E which explains the value conversion. Triple E offers the basic law his dynamic view on innovation. All these elements—innovative entrepreneurship, production factors, technical, nontechnical innovation, emulation, Triple E, innovation context, and value conversion—are brought together in a coherent theory. This theory is called strategic innovation theory (SIT) (van Someren 1991, 2005). May be even more important, SIT has been used in practice at small and big organizations in the private and public sector to create, find, and implement strategic innovations.

The strategic innovations are the mirror of the history. Green China is trying to follow a strategic innovation path in the future but is missing crucial experience with these transformation processes. The West has created strategic innovations in the last three growth waves but missing the understanding of the new challenges of Green China.

The core issue is about whether there will be a European economy will be able to coexist with rising Chinese economy or whether it will be dominated by a RMB-based economy. From all perspectives, whether it is policy, industry level, or corporate strategy, the rise of China leads to fundamental new thinking, behaving, decision making, and therefore innovating. This book introduces the possibilities of SIT as a scientific and practical instrument for China and the West to find a way to the future. The one which want to innovate strategically will see what future action is needed for any world player. In the next decade, it will be decided whether the outcome will be a clash of systems, trash of systems or cash by one of the systems, or adapted combination of systems.

Sustainability is not a single issue but an integral part of a range of innovations. Unfortunately, sustainability is still seen by the West as moral duty instead of a new growth direction. With regard to sustainability, SIT is not another management instrument or business process to follow but much more a core idea about society. Implementing sustainability is for most Western managers a huge problem, and

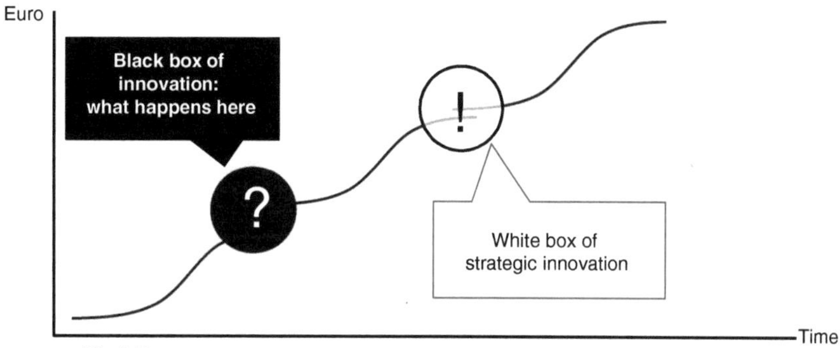

Black Box		White Box
- Technological trajectories/R&D	Content:	- Renewal: learning from history
- Revolutionary		- Economic value creation
- Once in a time		- Time dimensions
- For the happy few		- Growth and uniqueness
- Only for specialists	Creativity:	- Innovation of the business model
- Little knowledge about		- Sources and processes
- Luck and coincidence		- Knowledge creation
- Hard to manage		- Entrepreneurship
- Aspects of innovation	Context:	- Organisational and cultural context
		- Methodology of strategic innovation cycle
		- Mobilization of creativity and knowledge
		- Comprehensive view on innovation

Fig. 1.5 Strategic innovation: Know why, what and how. (Source: van Someren, 2005)

many CEOs, vice presidents, directors, division heads and partners, and MBA-trained managers do not know how to deal and implement sustainability. This shows that our idea about aim and organization of our society, the way we do business, and our behavior are not sufficient. These comprehensive strategic innovations together will lead to a sustainable society.

In this book, we will use the metaphor of the growth cycle in the form of an S-curve which can be applied on all levels and for all societies, industries, individual organizations, and single products or services. While dealing with sustainable policies and strategies in both the West and China of central and local governments, networks, and private and public organizations, we will demonstrate that the old world and market order will be replaced by a new one. The new world is being shaped right now, but at the moment of writing, we do not know how it will look like exactly. But based on our analysis of polices and business strategies in China and the West, we know which battle fields will decide the winners in the future. Figure 1.5 shows the growth cycle as a core instrument of the book.

The decade-long European dreams of growth and the American desire of a Sputnik moment are represented by Fig. 1.5. It is especially amazing that the ever innovative and creative US society has missed the mega challenge of sustainability and has no answer in the rise of China. For the EU, it is less amazing but even more necessary to really be able to create new earning capacity and withstand the rise of

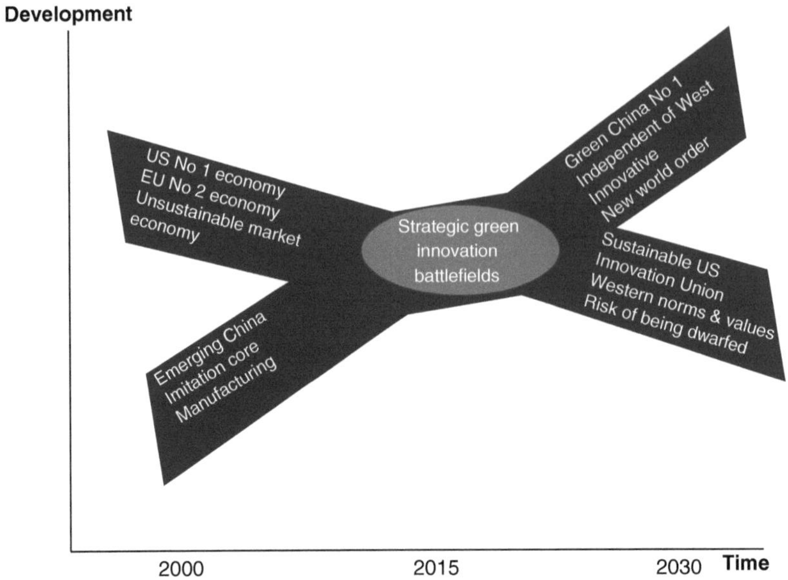

Fig. 1.6 EU and China on a crossroad

Green China. Both the challenge of sustainability and the rise of China are black boxes. The key challenge is to get insight into these black boxes and have full understanding of the inner workings. In fact, the USA and the EU are on a crossroad as represented in Fig. 1.6.

We are on a crossroad which requires new answers in the form of strategic innovations which will together lead to a new world order. This development from an old to a new curve represents the renewal process called strategic innovation. This process of strategic innovation takes place on all levels like single products, services, individual organizations, network of firms, industries, regions, nations, and continents. Hence, strategic innovations take place on micro, meso, and macro levels. If you understand the mechanism and rules how to transform your organization from the first curve to the second curve, you are able to survive in a future world.

However, Fig. 1.5 is a very simplified picture of the real world. There is never one big jump but many small steps, planned and unplanned, mutually dependent sometimes independent, which gain power and accumulate and reinforce each other. But the outcome is not only dependent on the future developments but also on existing history, bottlenecks, behavior, and world views as Fig. 1.7 illustrates.

The jump from the old to the new growth and development in Fig 1.7 will be decided by the outcome of ten battlefields. In summary, Green China comprises the following issues:

1. New world view: China wants to regain its self-denounced natural position being the center of the world. China aims to become number one in every respect.

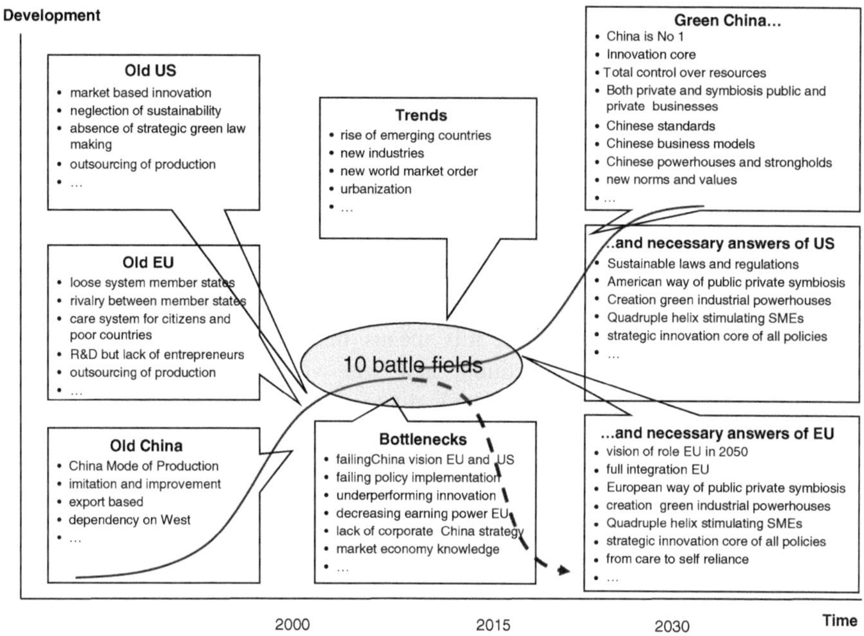

Fig. 1.7 From old to new world. (Source: based upon Van Someren, 2005, Strategische Innovationen, Gabler Verlag)

2. New political market economy will emerge. Green China implies a new world order with new rules for international relations and doing business. As a consequence, the Western distinction between macro level (e.g., securing resources, RMB), meso level (e.g., creating new industries, standards), and micro levels (e.g., selected firms and indigenous innovation) is blurred.
3. New market and corporate governance rules: A hybrid system of both Chinese informal networks of nongovernmental private firms as well as a symbiosis between private and public organizations will blur free market concepts. Due to current European policies and private strategies, the EU and China are on a crossroad and a battle for the level of coexistence has started.
4. Total control: China wants to become independent of the West especially technology and to have control over resources and value chains.
5. Sustainability: The core issue and challenge of the new growth curve is greening of industry and society. But with the rise of China as the dominant player of this new era, for the West, many more challenges than sustainability lie ahead due to above-mentioned game changing developments.
6. Emulation: China will develop its own strategic innovations based on a global approach combined with a new national innovation system. The name of the game is emulation meaning that a long-term strategy from imitation to independent innovation implies to learn from the West and overtake.

7. Chinese powerhouses: Chinese mega firms will enter global market with Chinese business models. New powerhouses will enter global market. In the EU, all public and private organizations on all levels need strategic innovations to give an answer on Green China in order to survive. This implies that formulating new business models is not sufficient but additional measures beyond the traditional corporate boundaries on all levels are necessary. New battlefields emerge. Individual private and public organizations need China strategies which go beyond outsourcing and market access for technology.
8. Chinese behavior. Asiatic–Chinese norms and values will enter market society influencing decision taking, strategic thinking, cooperation, and competition. From a Western perspective, this means being the best, overtake others, hardwork, cutthroat competition, and win–win means the Chinese win. Westerners will encounter paradoxical situations and ruthless behavior like a long period of cooperation and being kicked out as partner from one day to the other. Especially Europeans think equality and rules is the norm of relations between human beings. Chinese relations between human beings are based on power, hierarchy, and network background. In this respect, Westerners are not equal to Chinese; Chinese are better. Westerners will be respected till they become useless, for example, when knowledge has been transferred.
9. Nontechnical innovations: The key for becoming a true innovation state, the aim of the USA, the EU, and China, will be less dependent on technology and more dependent on nontechnical innovations. Examples are bottom-up innovation processes and private industries and cross-cultural management. All can learn from each other.
10. Redefining win–win: The Western mantra of win–win is different than the Chinese win–win. Only the ones who are able to get insight in each other's strategies, hidden agendas, and emulation process are able to survive and be able to create and sustain a countervailing power against each other. This is necessary because a unilateral domination of the world of any nation state is unlikely as long as the ability to renew itself and to take up challenges is inherent to societal behavior.

The West needs strategic answers to these groundbreaking changes that will change the business environment more and faster than the predicted climate change. Some speak about a system change, but we reject this kind of terminology. System change has the connotation of a mechanical process where all inputs, inner workings, and outputs are known and we only have to push the right button to implement a new system. The opposite is the case. Small and big steps taken under uncertainty on all levels somehow come together and form a new system. So everybody contributes but nobody has complete control. Changing the world is not a machine or a blueprint to be constructed. It is a process under uncertainty of human (mis)behavior, experiments, unexpected outcomes, success, and failure but each with the aim to achieve something.

1.8 Setup of the Book

China is a hot topic for the West for the wrong reasons. Some see China as the market paradise with unlimited chances. Some see China as the copycat with low-value economy. Some see China as factory of the world sustaining our welfare by producing cheap products. Some see China as a threat with dirty industry. Some see China as a superpower with international ambitions. Some see China as totalitarian regime with confusing mix of Marxism and capitalism. Some see China as the saving angel with a deep pocket. Some see China as the necessary evil with choking credit. The potential of a greening China is neglected. The focus of the book is not about political changes or possible ideal political orders but about the necessary strategic innovations to create a sustainable world and its value creation process. It will be shown that both China and the EU still have a lot of challenges to fulfill to attain the sustainable society. This will be shown with help of the strategic innovation theory which will address crucial aspects of the transformation process to a new world order.

Sustainability not only deals, although it is the initial trigger, with climate change, emissions, and other environmental issues, but also deals with welfare of the people and economic profit. In our view, sustainability together with the rise of China is part of other challenges which will renew our world order. Because of these profound changes sometimes accompanied by turmoil, it is necessary to connect sustainability to this larger context. The green sector targets all three aspects of sustainability. Facing the limitation of the growth based on old industry, China is forced to see the potential of the green sector and is trying to create one. With the growing market share in the green sector, the first sign of Chinese success is already showing.

A central shift may occur from Western thinking and acting toward Asiatic thinking and acting. Because it is expected that China will be leading, we concentrate on China and leave aside other Asiatic countries which can differ from each other as much as Western countries. A similar limitation will be made for the scope of the West. The West comprises European Union and its national member states and the USA. Western economic and management literature is dominated by North American literature and North American business cases. In this book, we will focus equally on the European Union and the USA for two reasons. First, sustainability has a longer tradition in the EU, but strategic innovation can be slowed down by absence of pan-European approach. Second, the USA, with innovation spirit as it showed in the past, may play a role in strategic innovations but still has to realize the essence of the real challenge of sustainability and the rise of China. Whatever future will emerge, the issue of strategic innovation will be the core of any development path and will appear and applied in different forms in the West and China.

The key message of this book is: Challenges like sustainability, the decline of the old economy, and the rise of BRIC countries are all signs showing that a new world order is coming. The new world order may have Green China as the center. The West neither sees the signs nor has an adequate strategic response on this biggest challenge of the last centuries. If there is not a strategic answer from the West, then another dream of Chinese will become true: to be the center of the world.

China is trying to generate a green economy to keep up the need of growth and to give answers on the question how to achieve a sustainable society. This process is not a rhetoric phrase or a hype but fits in its history and cultural roots. The foundations of the recent economic success will be used to transform China to a Green China. A strong symbiosis between Chinese government and individual Chinese firms forms the basis of a new sustainable growth curve. These new economic powerhouses will dominate domestic Chinese markets and stimulate the forming of national champions in the green sector. These new Chinese firms will conquer the world markets like the labor-intensive Chinese export industry do now.

The current answer of the West is insufficient. The sign of missing the ride of this green growth wave is already showing in, for example, the green energy sector with the invasion of Chinese solar panels in Europe. The main strategy of Western companies is to be present in Chinese markets mostly sacrificing heavy technology trade-offs to gain market share. The core argument is that presence in the largest future market is necessary in order to survive. It is a self-destructing strategy because the future dependency of revenues, profits, R&D, knowledge, and human resources will become dangerously dependent on China due to the size of operation scale of local earnings. Without alternative strategies, the Western economy will lose the vitality to sustain our welfare society.

Both policy makers on EU and national level and individual corporate organizations need to study the new Greening China mega trend and create necessary new strategies. The strategic innovation process helps the policy makers to find the right policy and the organizations to formulate the right business model. Most ideally, the public and the private sector join forces in the strategic innovation process to create a symbiotic growth strategy. The relevance of innovation, sustainable development, and rise of China as a dominant world player is meanwhile acknowledged by many, but in the West, we still lack clear strategies how to cope with these connected mega issues. Green China deals with these issues. This book gives insight into these necessary strategic innovations to build up a countervailing power in which both can win.

The core message leads to the following overview presented in Fig. 1.8 of content and connection between the chapters:

Chapter 1 has set the stage for the remainder of the book. Only a few historical developments have been addressed to show the relevance for understanding future developments of Green China. Differences in history, norms and values, and behavior cannot be denied.

In order to get more insight and knowledge of the future, Chap. 2 addresses the relevant megatrends for China. But they are no predictions or scenarios. It depends on your interpretation. In this respect, Chap. 2 also shows some very common Western reactions to the rise of China—China as a threat or opportunity. Both are not the proper reaction and another way has to be found. China is not the big unbeatable enemy of the West, but it is also not the ideal example of economic development and unlimited opportunities. Westerners need a different attitude and

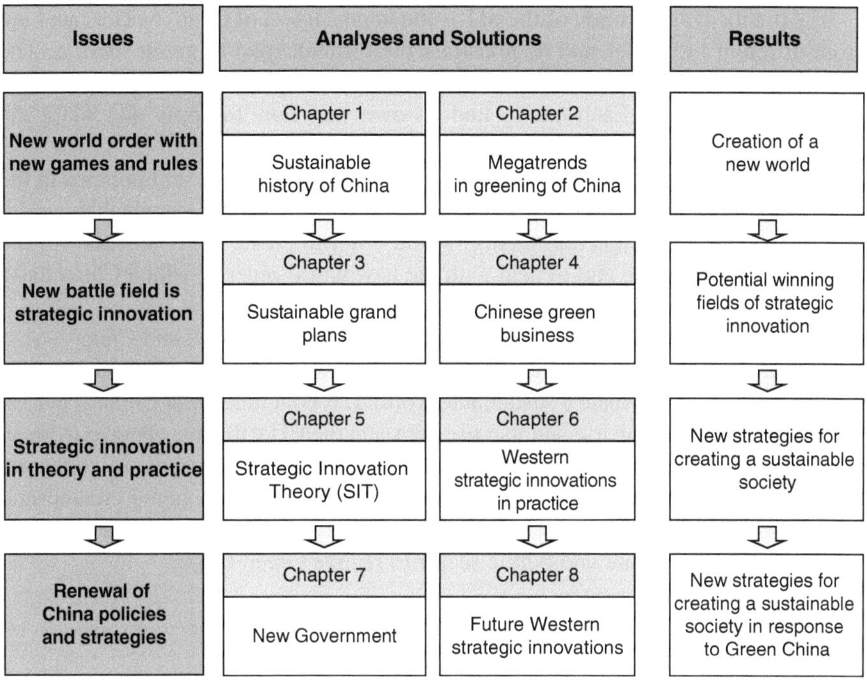

Fig. 1.8 Overview of content of chapters and their connection

judgement of China than the simple threat and opportunity thinking. For this purpose, we have to know what the strategic plans of China are.

Chapter 3 reveals some of the most important plans related to sustainable development. In particular, the Chinese five-year plans are discussed. They show the future investments and money flows which are of interest for Western businesses. But knowing the plans and having access to the relevant network is not enough as Chap. 4 demonstrates.

In Chap. 4, a few practical cases clearly reveal the missed opportunities of both political and business leaders due to a too Western view on Chinese demand and ways of doing business. These experiences are still actual and even hold for multinational companies active for decades in China. The selected cases showed the role of different factors of strategic innovation in China. Not only the hard part of business like technology is important, but also the soft part, especially leadership and people are more important. All the factors should be considered, adapted, and interlinked to be successful in the Chinese green sector. This gives us the directions of how to innovate our government and business. In relation to the core topic of this book, sustainability through Chinese eyes and strategic innovations are pivotal to future success and staying competitive with fast-growing Chinese rivals.

Chapter 5 elaborates on the highlights of the strategic innovation theory (SIT) which offers the methodology and practical instruments for creating sustainable corporation in and with Green China.

The practical application of the SIT is the main subject of Chap. 6. The cases are from different industries and demonstrate the difficult road to green success. The lessons learned prove that both the West and China still need to apply strategic innovation approach. Each has to find its own way how to apply and adapt the approach to its own circumstances.

In Chap. 7, the necessity of the creation of new governments and policies in the USA, EU, and also China are necessary to create a future sustainable world. Governments have to reinvent themselves in order to create the conditions for the new long growth curve and to deal with the new world order in which China has a prominent role.

In Chap. 8, it is demonstrated that all regions—the USA, EU, and China—need to implement strategic innovations in both private and public sector in order to be on an equal level and create a sustainable world. It is concluded that China is not yet the number one of the world and that the USA and the EU still have chances to be an equal rival of China in this century, but only when the necessary strategic innovations on governmental and corporate strategic level are being created and implemented. Also China is not ready yet; the country also needs strategic innovation in the private and public sector to realize Green China.

References

Friedman, T. L. (2005). The World Is Flat: A Brief History of the Twenty-first Century, Farrar, Strauss and Giroux, New York

Mahbubani, K. (2008). New Asian Hemisphere: The irresistible shift of global power to the east. Public Affairs. New York

Matsuno, H. (2009). China's environmental policy: its effectiveness and suggested approaches for Japanese firms, NRI Papers, No. 140, March 1, 2009.

Naisbitt, J., & Naisbitt, D. (2010). *China's megatrends: the 8 pillars of a new society*. New York: Harper Business.

Someren, T.C. R. van (1991). Innovatie, Emulatie en Tijd. De rol van de organisatorische vernieuwingen in het economische proces. Tinbergen Institute Research Series, no. 9, June 1991 (diss.).

Someren, T.C. R. van (2005 May). *Strategische Innovationen. So machen Sie Ihr Unternehmen einzigartig*. Wiesbaden, Germany: Gabler Verlag.

Tse, E. (2010). The China strategy: harnessing the power of the world's fastest-growing economy. Basic Books, New York

Wang, C. (2007). Chinese environmental law enforcement: current deficiencies and suggested reforms. *Vermont Journal of Environmental law, 8*, 159–193.

Xin, Q., & Honglin, L., (2009). China's environmental super ministry reform: background, challenges, and the future, environmental law institute®. Washington, DC. reprinted with permission from ELR®, http://www.eli.org, 1-800-433-5120.

Green Future of China 2

Golden Mountains and Black Holes: From Chinese Megatrends to a Radical New Management Agenda

What is the big picture of the rise of Green China? To answer this question, it is necessary to identify the megatrends in the Chinese society in order to have an idea in which direction it moves. But when dealing with China, there are always two sides of the coin. In Western eyes, they seem to be paradoxical, but coping with these several truths is a necessary ability when dealing with the future of China. Interpreting megatrends in a Chinese context is different from a Western context. The rise of China is accompanied with global investments in sustainability. The growth and development of China is driven and pushed forward by several groups within the Chinese society. The megatrends in China have far-reaching consequences for the West. Suddenly, several issues have to be solved at once by the Old and New World. The top management agenda for governments and corporate leaders is overloaded. A new world order is being created which will have a greater impact for the remainder of this century than the financial and debt crisis.

The sustainable future of China depends on solving the green paradox of China. On the one hand, the economic growth had tremendous detrimental ecological and social effects. Examples are CO_2 emissions and pollution of air and water. On the other hand, Chinese firms are worldwide front-runners in wind and solar energy. For the near future, this situation will not change very easily and not very quickly. The choice to open China and to choose the path of prosperity and wealth creation cannot be left very easily anymore. Large portions of the people of China have now experienced a better and wealthier life than previous generations. But even a larger portion of the population has not been levied in better and wealthier position. Economic growth has become a necessity to satisfy the needs and fulfill expectations of the large majority waiting for their share. It poses the Chinese leaders for the greatest challenge in recent history. Political lives and the legitimacy of Chinese government depend on synthesizing the paradox of economic growth

T.C.R. van Someren and S. van Someren-Wang, *Green China*,
Management for Professionals, DOI 10.1007/978-3-642-28810-4_2,

with ecological and social detrimental effects. The central government is aware of this and has taken policy measures to encounter the issue of sustainability and implementing Triple P performance criteria. On their turn, the actions to create a sustainable society by both Chinese government and business organizations put the West for an equal big challenge.

For the West, till now, China has been a part of their old world order in the form of the workshop of their value chains. By means of outsourcing, Western corporations, as market economics required, have used the Chinese cheap labor reservoir as an additional possibility to lower costs. From the perspective of strategic innovation, however, this short-term static strategy of cost savings now turns chances. There are two reasons the chances have turned. First, the buildup of the manufacturing base has created wealth and increased personal income levels of many Chinese households. The purchasing power increased. Slowly but steadily, a market for selling products has been developed. Second, the outsourcing enabled China to build up their manufacturing industry and to move on. The next step is to use the efficient manufacturing base to create sustainable products such as electrical cars, windmills, and solar panels. These two consequences of the dynamic market process pose the West for their own big challenges. One side of the coin is to earn money from the Chinese by selling products and services. It is the century-old Western dream of selling only one product to every Chinese, and the size of the population numbers makes you rich. The other side of the coin is that new Chinese rivals in old and new industries are becoming increasingly competitive. Therefore, in the near future, a complete new market situation will emerge. Because of the size and potential power of China, the reaction in the West boils down to two principal views. The first view only sees big opportunities and a huge new market ready to be taken by superior Western management. The second sees a big threat and is awaiting a tremendous tsunami of government-supported economic power crushing Western market-based organizations.

Or simply stated, China represents a golden mountain or a black hole. In the next sections, these two notions will be worked out in more detail. But to find out whether these extreme positions represent the real development in China, we use the method of megatrends to identify the big picture of the future of Green China.

2.1 Green Golden Mountains

The green golden mountains of China are the result of the size of China and its recent past and expected future development and growth. In the past decades, the focus was on economic growth and levying the Chinese people from poverty into the middle class. Annual GDP growth has been without exception around 8–12 %. In 2008, the nominal GDP per capita in China was just over $3,000 and in the USA nearly $47,000. So there is a long way to go. To a great extent, China's economic growth till now took place in coastal areas around megacities such as Shanghai, Shenzhen, and Guangzhou. The Go West strategy of central government tries to distribute wealth-generating regions throughout China. New large cities are planned.

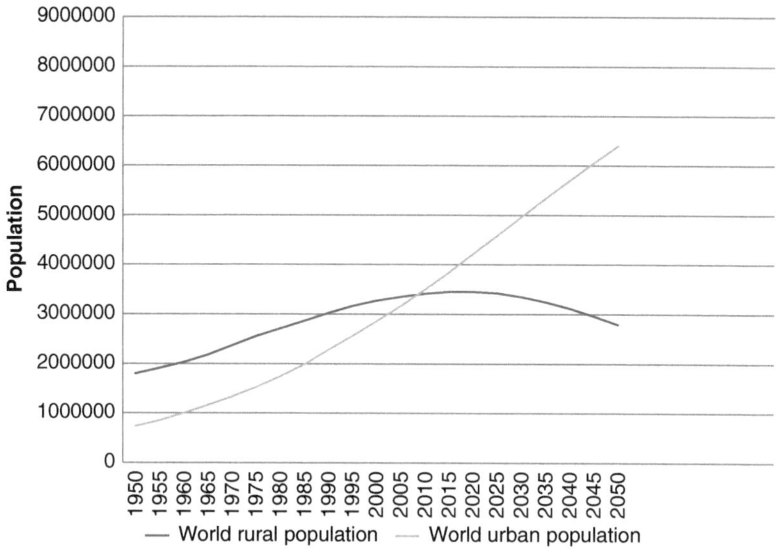

Fig. 2.1 World rural and urban population between 1950 and 2050. (Source: UN, The 2007 Revision)

The urbanization rate will definitely further increase. According to the UN in 2007, for the first time in human history, worldwide, more people lived in urban areas than in rural areas. Figure 2.1 shows this remarkable moment in human history.

Worldwide, the urbanization is expected to grow further. In developed countries, the urbanization rate is around 75 % (2005), and in China, the urbanization rate is 40 % (2005). China will probably reach the 75 % mark in 2050. In reality, this implies that hundreds of millions of people move from small towns to larger or mega cities. This growing urbanization is one of the megatrends which has consequences for many industries like mobility and transport, energy, water, agriculture, and services. Of course, many other (mega)trends like feminization and increasing complexity have an influence on the future of any society. For the purpose of Green China, in this chapter, only a few developments will be used to illustrate the possible golden green future of China.[1]

2.1.1 Green Car Mountain Rally

Only recently the steep growth of the Chinese car market has started. Already now, the larger cities in China are heavily congested with accompanying pollution and

[1] Main sources for data are The China Greentech Report 2009 and China's Policies and Actions for Addressing Climate Change, November 2011, Beijing, Information Office of the State Council The People's Republic of China.

health problems. Measured in 2008, in the USA, there are 80 passenger cars for every 100 people; in Japan, this number is 45 and in China 4. The absolute number of cars is expected to increase from 50 million in 2008 to 300 million in 2030. Despite limitations of car registration, the rising income level together with the growing urbanization will definitely force the car industry to reinvent itself and put forward strategic innovations. The golden mountain is turning into a green golden mountain. The stricter pollution and fuel consumption regulations pave the way for sustainable car concepts. Chinese automotive industry heavily supported by Chinese government is preparing for new energy vehicles such as hybrid and electrical cars. For Western companies like Daimler and BMW, plug-in does not only mean plug-in hybrids but strategic alliances with Chinese companies. The plug-in is their entrance to Chinese New Vehicles Program. In the eyes of the automotive industry, China is a golden mountain. Chinese manufacturers such as BYD, Chery, Geely, and Foton jump on the bandwagon of the new vehicle program.

The electrification of mobility is not limited to cars. China has the biggest market for electrical bikes and scooters. In 2008, it is estimated that approximately 65 million scooters and bicycles were driving around China. China is by far the biggest market for scooters and bikes, and it accounts for 90 % of all world sales. Accordingly, also the production has risen sharply but is getting competition from outside of China as well.

2.1.2 Green Energy-lighted Mountain

The history of developed countries proves that energy consumption rises with GDP and income levels. China has huge coals reserves; measured along current consumption, the known Chinese coal mountains are big enough to supply energy use for the next 400 years. Coal still provides 70 % of China's energy supply. It is no surprise that in recent years, it is estimated that every week, a coal firing plant has been opened. Furthermore, partly due to increased use of cars, the dependency on oil is increasing. Chinese import of oil is half of its seven million barrels daily consumption. At the same time, Chinese government knows that because of CO_2 and environmental detrimental effects, coal and oil are unsustainable. For this reason and because of rapidly rising demand, additional and renewable energy resources are not an option but a necessity. In the next decade, an energy mix seems unavoidable. All the big oil companies come up with the scenario that in the long term, 2030 or even 2050, an energy mix is the only option. For example, BP in their Energy Outlook 2030 assumes a share of fossil fuels (coal, gas, and oil) of approximately 75 % and the nonfossil fuel (hydro, nuclear, and renewables) of 25 % in 2030. Moreover, the energy intensity per tonne of oil equivalent (toe) of GDP will reduce.

Great opportunities exist for renewable energy sources such as hydropower, wind, tidal waves, biomass and solar. Other energy sources are natural gas including shale gas and nuclear energy. It depends on your view whether nuclear is regarded to be sustainable or not. The nuclear waste cannot be denied and remains a big issue, and after Fukushima, also safety is not under complete control. For the Chinese government, carbon capture and storage (CCS) is a very important technology in order to get grip on CO_2 emissions from coal, gas, and oil-firing plants. CCS is the Golden Gas Mountain and a road to low carbon economy. It offers a multibillion market. Western companies like French Alstom see golden CCS mining opportunities. In the last decade, Chinese firms in the wind and solar industry made a rocket start from scratch to world market leadership. In 2007, Chinese photovoltaic (PV) manufacturers had a worldwide market share of 30 %. The share of the total Chinese production of PV cells and PV modules for export is about 90 % whereas only 4 % is used for domestic consumption. The golden Sun Mountain attracted many Chinese entrepreneurs but also local governments trying to establish their own Golden Solar Valleys. Consequently, the production capacity could not match the demand and installation capacity. The overcapacity resulted in piles of PV solar cells. The solution was export, and around 95 % of total Chinese production is sold in world markets. Especially the German market with their long-term subsidies is an attractive market. Suntech Power Holdings, Yingli Green Energy Holdings, and Trina Solar belong to biggest manufacturers and are already active in the EU and the USA. With the new plans for renewable energy, large sums of central government are available to support and develop this industry. China plans to get 20 % of its energy demand from renewable resources in 2020. Hence, domestic demand for solar energy is stimulated. In the future also, Concentrated Solar Power (CSP) will contribute to the energy mix. But CSP is in China relatively in its infancy and far behind leading countries like Spain and the USA.

2.1.3 Golden Water Valleys

For any country, water is becoming more and more a pivotal resource. Water is not only necessary for human consumption but also for industrial use such as in food industry, power generation, and water-based paint shops in manufacturing. Access to water is as important as oil. China has several problems in the area of water supply and drinking water. Moreover, the South of China compared to the North possesses around 80 % of the water resources. According to the World Bank, between 2010 and 2050, the availability of water resources per capita in Northern and Southern China river basins will decline with 17 %. Droughts, disappearing forest, climate change, and intensified human use contributed to changing supply. Very important in this respect is the inefficient use of water for industrial purposes. For example, the agricultural sector consumes about two third of all water consumption. And on top of that, due to inefficiencies, only near half of the water is actually used for growing crops.

Besides water scarcity, China also has to cope with water pollution. It is estimated that approximately 70 % of all rivers are polluted to grade IV indicating that water is unsuitable for human use. The main cause is industrial activities without proper measures against pollution or failing sanitation. This holds from many industries alongside rivers such as agriculture, manufacturing, paper mills, and chemical industry. Annually, about 50 million tons of wastewater is discharged in Chinese rivers, lakes, and other reservoirs.

China has formulated several policies to cope with the above-mentioned problems in the water sector. In November 2011, the government announced various plans such as National Comprehensive Plan for Water Resources, Seven Major River Basins' Flood Control Plan, National Mountain Torrent Disaster Prevention and Control Plan, National Plan to Guarantee the Safe Supply of Drinking Water to Urban Dwellers, and National Plan for the Eco-protection of Major Rivers and Lakes. New laws are issued such as Water and Soil Conservation Law. New policies are issued for the development, utilization, conservation, protection of water resources, erosion control, waterborne health diseases, and irrigation. One of the aims is to guarantee supply of safe drinking water to 210 million rural residents.

2.1.4 Golden Food

The paradox of food is that on the one hand, the Chinese are obsessive about food and on the other hand, food safety is sometimes a very big problem. On the long term, the Chinese demand for food exceeds the supply. Therefore, besides water, the Chinese agro-food sector offers a huge challenge. Agro-food products are another golden mountain. May a better description would be another Golden Himalaya is to be seen on the horizon.

The market for agricultural products in China has the dimension of a Golden Food Himalaya measured in absolute numbers as well as world market share. China has about half of the world market share in the production of the beloved vegetables and pork. Moreover, it is expected that demand increase in China (15–20 % per year) will surpass supply increase, and the gap between the two has to be filled in the future, especially around metropolitan cities like Shanghai. This gap is one of the driving forces for introducing innovative agro-food production concepts. One of the reasons why it is difficult to meet the demand is that the supply is limited by the natural conditions of China. Only 13 % of China's soil is suitable for agricultures (van Someren and Nijhof 2010).

These harsh conditions and the increasing gap require new solutions, and none will be excluded. Especially, the biotechnology sector has enormous opportunities to solve the gap between demand and supply. Additionally, also the existing organizations in the agro-food have to implement strategic innovations. However, Chinese SMEs account for 80 % of China's agricultural production. Bigger commercial multinational agro-food companies are rather the exception but a few have world-class size. For example, the Henan Shineway Group has an annual turnover of 1.6 billion euros in the pig industry and Delisi Group has an annual turnover of 362 million euros in the cattle industry. With regard to the position in the value chain, it is interesting to note that in China, only 30 % of the food is processed. Compared to agro-food champion the Netherlands where 80 % of the food is processed, the Chinese share in food processing is rather bleak. But it represents a golden mountain.

In China's Policies and Actions for Addressing Climate Change (November 2011), the following policies and measures are announced: "China strives to consolidate farmland and water conservancy infrastructure, raise the overall production capacity of agriculture, encourage large-scale construction of standard farmland with stable yields despite drought or flood, support major irrigation areas to continue building supporting and big irrigation and drainage pump stations, expand the irrigation area, improve the irrigation efficiency and spread water-conservation technology, and carry out agricultural water pricing reform and pilot water-conservation renovations of farmland ditches, and enhance the nation's capacity to resist disasters. It has built and improved agricultural meteorological monitoring and early-warning systems. China also carries out research into and cultivates stress-resistant varieties of seeds with high yield potential and high quality and resistance to drought, water logging, high temperature, diseases and pests, and expands the growing areas of super strains of crops. It has further increased subsidies for superior strains of crops, and accelerated integration of the cultivation, reproduction and spreading of such superior strains. Currently, more than 95 percent of the farmland nationwide is sown with superior strains of major crops, which contributes 40 percent to the grain output increase" and "It has dispatched agricultural technical experts, trained agricultural technical personnel and improved Africa's ability to realize food security. It attaches importance to cooperation in human resource development, and has carried out 85 foreign aid training programs."

On paper, these policies are very promising, but they require an agro-food sector able to create the necessary innovative changes.

2.1.5 Market Value of the Golden Mountains

The several worldwide crises skyrocketed gold prices. Sustainability issues in China skyrocket the total worth of Green Gold market in China. It is estimated that already only relevant sustainable issues taken together in all industrial sectors represent a total market volume up to $1 trillion per year. Which real businessman or businesswoman can resist such an opportunity of earning money and personally becoming rich and do something for the world? No one.

Europe has emulated the world by paying attention to environmental issues in an early stage. Governmental policies; strict regulations on, for example, air emissions, water pollution, and soil protection; NGOs, and pressure groups together with some innovative entrepreneurs ignited a market for sustainable products and services. Europe had a head start and the green mountains are in the eyes of many entrepreneur and manager really gold. The timing is now right to earn money from China, they think. And they act by traveling to China and try to get involved in the many projects. Some of them are successful. But for how long?

But China is not at a standstill. Even in some cases, China emulated Europe's top of the bill companies in the wind and solar sector. Therefore, it is time to ask oneself whether the green mountains are really made of massive gold or whether it is a black hole.

2.2 The Chinese Black Holes and Twin Stars

Are there no green golden mountains, but are we attracted by a black hole? Every galaxy has one or more black holes. Compared to other nations, China itself is a galaxy, and there are many black holes. There are several factors which could die out the energy of any star in the galaxy. Four of them are manifest and substantial (van Someren and Someren-Wang 2011).

First, the housing bubble which already has brought construction companies, house owner, local government, banks, and stock trader into trouble. The decreasing prices for houses lead to meanwhile acknowledgeable financial problems.

Second, the number of bankruptcies in low-value industries like toys and clothing is increasing. The main general causes are rising labor costs, reduced demand, and less private loans. The chosen solutions like outsourcing outside of China and increased use of robotics are detrimental for Chinese labor creation and are part of upcoming social unrest.

Third, the fear for bad bank loans increases. Estimates which indicate that a quarter or even more of the bank loans will be irrecoverable are no exception anymore. Not only banks are to be blamed but also irresponsible loans and unlimited urge for expansion by local governments contribute to increased financial risk. Meanwhile, the Chinese government has taken measures with regard to bank reserves for banks.

Fourth, the Chinese government possesses a national currency reserve of more than 3,000 billion euro. But the Chinese government is now more prudent with spending and investing out of fear for a Chinese crisis needing substantial financial power for a forceful counterattack. Premier Wen has already announced that for prestigious projects such as the electrical car, no unlimited pots of money will be available.

Some of the black holes are disguised through interstellar nebula as golden mountains. Doing business in China is for many Western companies a challenge, even for those who have been involved in China for decades such as Volkswagen, Philips, or Siemens. New policies, new regulations, different interpretations of regulations in different time periods or in different regions, rapidly changing market conditions, new customer demand, and unexpected behavior from business partners make Chinese markets volatile, unpredictable, uncertain, and hard to make money in. There is no one single market but various markets which can differ per city, region, and province. They all need their specific attention of market needs, customer behavior, different regulations and governmental attitude, and decision making. Furthermore, also Chinese like golden mountains. Even worse, almost every Chinese is an entrepreneur, and they all want to have their own company. Even worse, every Chinese entrepreneur wants to have the greatest and biggest company of the world dominating anyone else. The last decades, the European

Commission and European national governments have desperately formulated different policies to stimulate entrepreneurship, without success. So for the West, the big challenge is as follows: How to fight 1.5 billion Green Dragons?

In all industries, the Chinese market is very dynamic, but the downside is that almost in every sector, overcapacity is the result. Not only the copycat behavior of many entrepreneurs is the cause but also the local governments who all want to have their own golden mountain. Overcapacity means low profit margins or even losses. The Chinese solution is manifold. From a Western perspective, two alternatives are relevant. The first solution by Chinese government is forced merger in an industry in order to reduce overcapacity and the opportunity to create power houses able to dominate world markets. These Green Dragons are used to fight foreign companies penetrating Chinese markets. The second solution is export supported by Chinese government. The case of PV solar panels and modules demonstrates this market behavior. For example, the overcapacity in wind energy sector intensifies competition, and a shakeout in the industry is unavoidable. Another aspect is the succession of new technologies which make older generations of technology obsolete. Illustrations of this effect are the first- and second-generation solar panels and led lamps.

In nearly every circumstance, Western organizations have to cooperate with Chinese organizations in order to be able to do some business. Central Chinese government has earmarked several industries as strategic such as banking and automotive industry. Western companies need not only a Chinese alliance partner, and they are only allowed to have a minority stake in the joint venture. Meanwhile, every Western manager knows what the purpose of these alliances is. For the Chinese, only one thing counts: the access to Western design methods, R&D, patents, technology, and managerial knowledge of branding, marketing, and distribution. For Western organizations, it is a trade-off between market access and technology transfer. Therefore, a China strategy with the adequate strategic innovations is necessary. The golden green mountains can indeed become black holes. The huge demand and opportunities represent an irresistible opportunity requiring a never-ending stream of investments in resources and time to build up your Chinese market organization. Each new involvement in Chinese market, such as establishing new factories and opening new regional markets, makes you believe to come a step closer to conquering Chinese markets. But with every new investment, you come closer to the black hole. There is a boundary between your Chinese organization and the black hole where you still decide to turn back and move in another direction. If you are too late, you pass the death zone and no return is possible.

The black hole for pollution generators is in some cases nearing the death zone, and in other cases, this border has already been passed. The production sector was attracted by favorable conditions to set up factories in China in the 1980s and 1990s of the last century. It is even the case in some underdeveloped regions in China now. Environmental regulations and the implementation were mild for those who could bring money and jobs. But they are no longer welcomed once a region is rich enough to realize the consequences of pollution. China wants to be more than the factory of the world (van Someren and van Someren-Wang 2009). Those who cannot see this trend and continue to produce in an unsustainable way will be forced to close down or out of the market. There is a very big and powerful black hole for them. Examples are the creation of eco-cities and eco-economic zones, the environmental friendly paint market, the reorganization of the city heating, power plants, steel industry, leader industry, cloth industry, and toy industry which all will have sustainable characteristics.

In nonstrategic industries such as retail, most Western companies need a partner to get access to the distribution network. In this case, the problem is not to get or keep your majority stake but to increase your total market share for China as a whole. The cutthroat competition is really economically murderous. In these cases, business partners are not black holes but twin stars in which the Chinese star sucks the Western star to death. The stream of gases full of Western knowledge, patents, and experience is the lifeline of the growing Chinese star. Also here, the transfer can take months but also years. Again, at the end, the Chinese partner wants to be the winner, and they will do everything for that aim. Your Chinese business partner will show you great respect, offer you great unmatched hospitality, and honor you and your company, but the next day, the brother or nephew of your business partner imitates your company at the other side of the street. How can their respect and hospitality suddenly turn into your economic enemy? Because they regard themselves best; they want to become the number one, and in the eyes of Chinese, it is your own wrong strategy or naivety if you do understand their behavior. It is you or me. Black holes are unforgiveable.

Chinese markets can be a continuous promise of profit and investments after investments are required to keep up with developments and opportunities. Sun Tze said, "Timing, positions and manpower harmony are the most important conditions to win a war." For Westerners, those are still important conditions to be successful in China. Firstly, you have to enter or leave China at the right time; retreat on time is considered the most important tactic in the Chinese 36 stratagems. If you come too near to the black hole, it will be too late to escape. Secondly, you have to stay alert of the national and regional policy development all the time and to decide your own position in the big picture of Green China. When you cannot stay ahead, you will have to find another direction. Thirdly, keeping harmony with your stakeholders is also important in China. Not only the shareholders or the market are important, the people in and around of the company are more important. Otherwise, in the long run, you will lose the status of part of twin stars and cannot fight the power of the black hole. Therefore, in case of a black hole, know how far you can go and choose the right timing to turn around. In case of the twin star, if you have to team like a twin star, be sure to be and stay the bigger one. True black holes and twin stars need strategic innovations and China strategy.

2.3 Megatrends

Worldwide megatrends give an indication in which direction the future world is developing. But there is no unequivocal approach to megatrends. In Table 2.1, a few different perspectives of megatrends are presented.

Table 2.1 Megatrends by different parties

	Copenhagen Institute for Future Studies	Frost and Sullivan	Laurence Smith	Dutch fire department
1	Aging	Urbanization	Demographics	Sustainability
2	Globalization	Social trends	Increasing use of earth's resources, services, and gene pool	From natural risk perception to total safety control
3	Technological development	Economy	Globalization	From hierarchy to networks
4	Biotechnology	Technology	Climate change	The individual in the center
5	Nanotechnology	Energy		From life time relations to more complex (labor) relations
6	Prosperity	E-mobility		Toward a new societal order
7	Individualization	Infrastructure		Faster technological development
8	Commercialization	Business		Emergence of contrasts in society
9	Health and environment	Health and wellness		Decreasing Western dominance
10	Acceleration	Smart factory		
11	Networking organization	Industry-specific trends		
12	Urbanization			

There is no absolute certainty about these trends, but some of them are really robust. Table 2.1 demonstrates the different approaches toward megatrends, a very condensed approach of four versus the summing up of more than ten megatrends. It also depends if a general approach is preferred applicable on every industry or an industry-specific approach. The latter requires the translation of general trends into industry-specific long-term developments including industry-specific trends. Therefore, it depends for which purpose the megatrends are used. General megatrends for the coming decades boil down to demographics, technology, new forms of organization and governance, sustainability, power shifts, and economic, social, and political changes.

It will be no surprise that megatrends are very important for formulating your corporate strategy. However, when creating strategic innovations, megatrends are only one of the many ingredients as will be explained in Chap. 5.

For now, the most important issue with regard to megatrends is as follows: What are we doing with these trends and what is corporate and policy response? It is remarkable how both corporate leaders and policy makers are aware of trends but do not react. For example, it is known for a long time that aging is characterizing several Western countries, but hardly any specific products or services are developed. On the contrary, governmental services or cable services for connecting TV are increasingly managed by Internet only, for most elderly people a hocus pocus. Similar, managers in the automotive industry, despite feminization, cars are developed for the male segment and specific female requirements such as a simple dashboard with only the necessary control buttons instead of a full loaded dashboard are hard to find.

For the purpose of this book, we will focus on some major developments in China. The academic father of megatrends, John Naisbitt, coauthored with his wife Doris, published a study of megatrends specific for China. According to the Naisbitt and Naisbitt (2010), there are eight pillars of the future Chinese society: the emancipation of the mind, balancing top-down and bottom-up, framing the forest and letting the trees grow, crossing the river by feeling for stones, artistic and intellectual ferment, joining the world, freedom and fairness, and from Olympic gold medals to Nobel Prizes. Their core message is that China is developing a new social and economic model called vertical democracy. The main task of the vertical democracy is to balance bottom-up and top-down forces. The basis for the major change in China is liberalization from a Maoist ideological mindset toward a mixture of freedom and control. The latter released the entrepreneurial spirit in China. The continuity of Chinese leadership enabled to make long-term plans. The Naisbitt's go even a step further by stating that the continuity in Chinese leadership in fact enables a continuous reform process in big corporation style.

How to deal with the golden mountains, valleys, black holes, and twin stars? A simplification might be helpful to make our point about the necessary intermingling of different topics. We use the metaphor of the rainbow.

Green golden mountain resembles the maximum optimistic variant of greening of Chinese society coupled with earning maximum profits for Western firms. The black hole and twin star variants represent the maximum pessimistic variant in which only the Chinese earn money at maximum cost of Western firms. Both the black hole and green gold are not only imaginary possible future states without any sense of reality. On the contrary, both really exist in the Chinese universe and did already happen. In some cases of outsourcing, Western firms maximized their profits without transferring any knowledge or technology. The opposite also occurred; the main inventors of the Maglev train technology, Siemens and Thyssen Krupp, only commercialized one short track between Shanghai Pudong and Shanghai Airport. All technology and knowledge of this fast magnetic train technology is in the hands of the Chinese. Therefore, any situation can happen in future. For this reason, we use the metaphor of the rainbow from Chap. 1 and apply it on the megatrends. In the next section, the focus will be on the megatrends in China.

2.4 The Colorful Future of China

The general worldwide megatrends are valid for China as well. But as demonstrated in the previous chapter, China often shows two sides of the coin which appear paradoxical in Western eyes. The other side of the coin with regard to megatrends in

China is not only the existence of opposing trends but also another interpretation of and dealing with megatrends.

The core message is that the synthesis of the paradoxes will lead to other solutions and other directions of future development. Already here, we encounter the difference between Western and Chinese thinking and behaving when creating a sustainable future. The different context, institutions, behavior, decisions, and so on lead to different appliance of megatrends in creating strategic innovations.

The history of China shows that its development has always been characterized like moving back and forth between the opposites of paradoxes. For example, for the elderly generation, saving money was common use, but for the current younger generation, it is less important. But the current behavior does not mean the saving behavior has disappeared; it will come back in time.

In the past, China has had periods of opening and closing the country for foreign influences and contacts. It is hard to imagine that nowadays open China will be closed again. But it depends on the circumstances. It is like in contemporary Europe where the outbreak of wars between European countries seems to be impossible and something of the past. But the tensions around euro and debt crisis show that old sentiments between countries and dwindling solidarity regain power very quickly. China has its own problems such as the many minorities like in different parts of the country or rising income inequality or the increasing transparency due to new social media. When these kinds of development become more powerful, a less open China is not unimaginable.

The different and changing shades of the rainbow of megatrends culminate in a colorful future of China. The methodology of DESTEP (Demographic, Economic, Social, Technological, Ecological and Political) is used to categorize the major developments in China. Therefore, the China insight view gives often a different picture than the self-proclaimed Western trend watchers. Only a few examples are given and especially some deviations from worldwide and accepted China trends are discussed and highlighted. The purpose is to be aware of the China watchers and to train yourself in your own judgment and to get acquainted with the fact that the clocks in China often tick different.

2.4.1 Demographical Trends

Only a few major trends are mentioned here, but they all show the different long-term trends with regard to demographics.

With regard to demographics, the aging problem in China is one of the big issues for the near future. The one-child policy is the main cause of the aging problem in the near future of China. In the West, the USA has no aging problem due to immigration. In the EU, aging is a big issue because of low birth rates in relation to mortality rates. Moreover, the recent years in many member states of the EU, immigration from non-EU states is a hot issue. Immigration could partly solve the aging problem, but the political climate is not favoring these kind of solutions. As a result, European countries have to take other measures to be able to maintain living standards and payment schemes to retired people. Most common is slightly raising the retirement age from 65 to around 67. In China, the aging problem will take another direction. For Chinese, it has always been normal to work your whole life, and the aging problem will be solved in this way. Chinese people will work longer or even lifetime to support their elderly family members. But for the one-child generation, the burden is higher than in previous generations. The family restructuring process, known as the 4-2-1 (one child, two parents, and four grandparents) problem, has the potential to undermine old family traditions, and new agreements on family life have to be sought.

Another Chinese megatrend is the power shifting between different age groups and classes caused by the shifting of the group size. For example, the green people, those are the young ones, who will be baring the burden of the consequences of graying society and unsustainable economic behaviors, may grow into a more powerful group. They are able to react like this due to higher education levels and the demand for higher educated people. The gray ones, the elderly people, will become dark gray (young elderly) and work harder and longer to maintain their position. Lifetime work will not be an exception in order to be able to take care of income for family members and secure a pension. The blue-collar workers will become scarce and better paid compared to first generation. The white-collar workers have to survive by getting a well-paid job at one of the rising stars in public- or private-owned companies or in public services. However, because of global pressure, the number of jobs is not endless. Therefore, more white-collar workers will be forced to establish their own business to survive, which is a good development for the economic creativity. Some of them will become black horses. Black horses are self-made people with great wealth and often power. Black horses will complement red princes. The past decades, the so-called red princes, the party leaders, and often their close family members will not be replaced but at least complemented by black horses.

What about the feminization megatrend? Traditionally, and based on Confucius' hierarchical thinking, boys are preferred above girls. That does not mean Chinese women have no chances. On the contrary, again based on Confucianism, highly educated people are respected, and therefore, highly educated women are indeed respected. A part of the female population will take their chances and make career and get top jobs. But as soon as the group of upper middle class and top class in society becomes bigger, the necessity for women to work will reduce. As a result, the roles of housewife or just enjoying life and spending husband's money will for a part emerge. A mixture of housewives and top managers will exist.

The above-mentioned developments will have an impact on the ongoing urbanization. A mixture of different shades of people will characterize the major cities. They will have a demand for all kind of different services ranging from agriculture food products, mobility, health, financial planning, luxury goods, to housekeeping.

In conformity with global trends, the urbanization in China is an ongoing process for decades. Figure 2.2 clearly shows this megatrend.

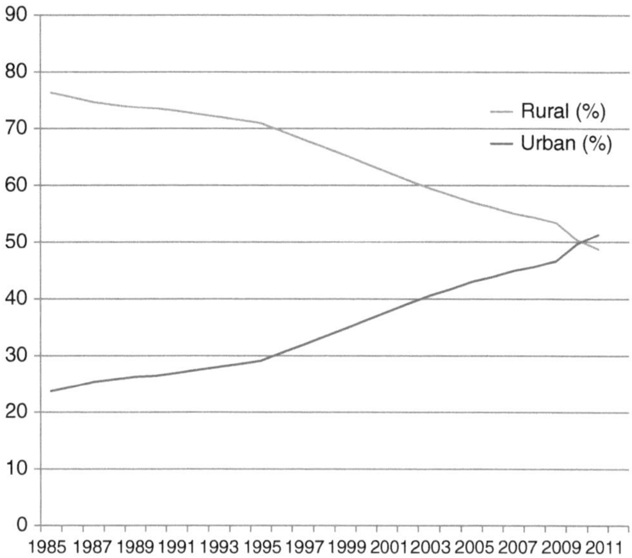

Fig. 2.2 Chinese rural and urban population between 1985 and 2010. (Source: Chinability and The national Bureau of Statistics of PRC)

In Fig. 2.2, it can be seen that as of 2011 in China, more people are living in urban than in rural areas. Probably, the worldwide trend of urbanization will be continued. For 2030, the UN expects the partitioning of rural and urban population to be 38.1 % and 61.9 %, respectively (UN). McKinsey expects the contribution of cities to GDP to grow from 75 % to 95 % in 2025 (Hefele and Bade 2011). There is a possibility that in China, a major counter trend toward de-urbanization will come up. The reasons are the unhealthy living conditions and the enormous problems with mobility. Both lead to substantial loss of economic and social welfare. It remains to be seen how long this will be acceptable for urban population. As long as the cities remain economically more attractive due to higher wages and educated people have more chances in urban areas, the urbanization will continue till new opportunities in rural areas arise.

Demographics with regard to Green China need also a global perspective. The Chinese people are the most dispersed people of the world. Exact numbers are not known, but it is estimated that the worldwide Chinese community counts around five million people. This migration has historical reasons going back till the nineteenth century when overpopulation, wars, and famine but also good prospects in the Western world stimulated Chinese to try their luck. Because the Chinese culture emphasizes education and being successful, many overseas Chinese grabbed the chances in the free economy and become successful. In Asia, the Chinese live in countries like Indonesia, Malaysia, Thailand, Vietnam, and Singapore. Although they only are 6 % of total local population, they possess 70 % of total held private and corporate capital (Kleining 2008). Since the opening up policy, China faced a serious brain-drain problem. The most brilliant ones went abroad by studying abroad. In the USA, by 2006, about three million Chinese expats are studying, working, and doing business. In European countries, Chinese do show same migration behavior.

The Chinese government is trying to get the brightest ones to get back to their home country. It remains to be seen whether this attempt of reversal to bring back lost sons and daughters will be successful. In the eyes of Chinese, they are bananas, yellow outside and white inside; only those who can use the best of the two worlds can be successful. Many may not suit the real Chinese society anymore. Moreover, going back to overcrowded dirty cities with intense competition and leaving nice careers and location elsewhere needs tough considerations. Even if they stay abroad, they will still contribute to the development of Chinese economy by doing business with their families and friends. The Chinese network is one of the reasons that Chinese export could flourish.

2.4.2 Economic Trends

The biggest economic megatrend will be the transformation from manufacturing economy toward an economy based on innovation. The result will be that other links like R&D, design, marketing, sales and distribution, and recycling in the value chain need to be developed. The main background is rising labor costs, shortage of skilled labor, and the slow appreciation of the Chinese currency RMB. China will stimulate both domestic consumption and foreign direct investments. The latter is necessary to secure natural resources but also to create markets for Chinese products and services. The higher value of RMB will make it more attractive to spend money on domestic or even imported products. The export of low added value will become less important in the GDP growth; export policy of China will be diverse toward exporting high added value products to developed countries and low added value products to underdeveloped countries.

The steep rising income levels can be seen form Fig. 2.3. Figure 2.3 also shows the increasing gap between rural and urban income levels.

The net income levels in urban levels are substantially higher than in rural areas, and it explains the increasing urbanization. But as observed before, the urban high-income groups pay a high price with unhealthy air pollution and immobility.

China's future economic growth will be a combination of exploring global and regional economic interest and stimulating national economic potentials and resources (natural resources like rare earth, human resources, and capital resources). Furthermore, market pull and high-tech-based push economic policies will be more important than only capital based. Both the market pull and high-tech push will be dependent on innovations. China's strategy for becoming a superpower will be dominantly based on economic power and less on military power. Chinese military budgets have risen and will so in the near future. But the main instrument for conquering markets will be economic power and the military power as a sign of regained world leadership. Both current industries like agro-food and

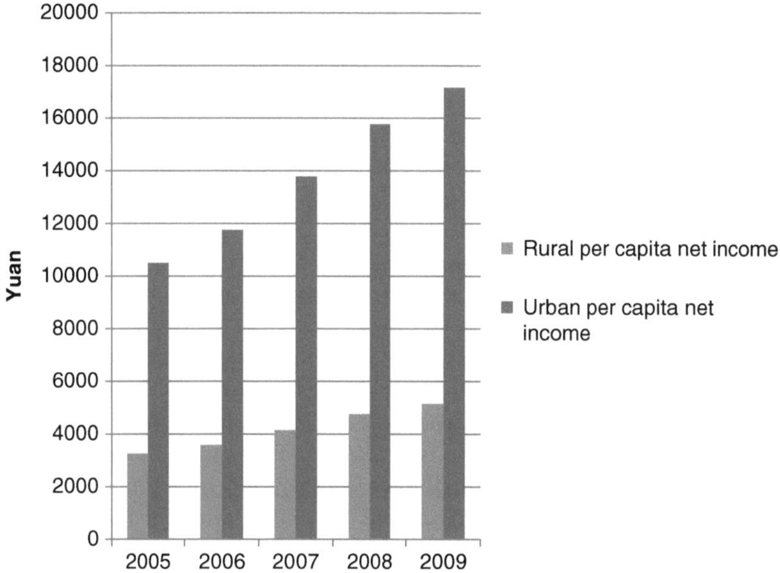

Fig. 2.3 Rising income levels and widening income gap in rural and urban areas. (Source: China daily 020310 / National Bureau of Statistcs of PRC)

telecommunications and the new industries like renewables, eco-mobility, biotech, and nanotech will be part of the future economic power. In both old and new industries, strategic innovations will transform the business.

The megatrends of urbanization, rising income levels, and sustainability stimulate metropolitan agriculture concepts. Local food production fits to shorter value chains or even value circles. Similarly, local energy production sometimes combined with local water production depending on the local situation will make the megatrend toward localization increasingly relevant. The expectations of McKinsey and United Nations about ever increasing urbanization is based on the assumption of a continuing dichotomized population living in either the (mega)city or the rural area.

This simple duality assumption can be doubted for one major reason. Based on the theory of strategic innovation, the bottlenecks of the (mega)cities will lead to a reinvention of urban–rural areas (van Someren 2005a; van Someren and Nijhof 2010). Strategic innovations will challenge the rural–urban dichotomy thinking of policy makers and business leaders and lead to new forms of regional growth areas. These regional growth areas, the border line between city and rural areas, will disappear, and new eco-city–regions are the future. Furthermore, together with the rejection of the dichotomy rural–urban and its replacement by regional growth poles, new opportunities emerge.

In the coming years, a coexistence of high-tech industry, low-tech industry, and no-tech industries such as services will characterize the Chinese economy. A shift of low added value activities will be outsourced to other countries in Asia or other regions in the world. The creation of an innovative economy is a necessity to deal with rising labor costs and solving the sustainability issues with new solutions. This is also what the Chinese central government tries to achieve: a turnaround from low cost, low tech, high volume, low added value to high tech and high added value economy. The current Chinese economy does not only consist of cheap labor-based manufacturing like clothing and assembling computers but also imitated-improved middle-high-tech products like solar panels and high-speed trains. In these industries, the first-generation entrepreneurs created the first-generation Chinese champions and sometimes multinationals. But their growth model was based on low cost and copycat technology.

The shift toward high-tech sector is not just another step or a next level. Despite their education and entrepreneurial spirit, it will be extremely difficult to achieve the desired high-tech knowledge economy. The Chinese first-generation companies will experience difficult times for two reasons. First, governmental focus and sometimes support will shift to high-tech sectors, and investing in old sectors might be the Chinese black holes. Secondly, becoming a true innovative organization or economy is extremely difficult as the experiences in the EU and the USA show. Competition based on innovation is completely different than low-cost rivalry. Moreover, organizational innovation like bottom-up processes will be key to being successful, and this is one of the big leaps forward the Chinese have to take. Therefore, strategic innovations and not technology will dominate future trends toward a sustainable society.

The economic focus in the past decades was on building an infrastructure like bridges and railroads, housing, and basic industries like chemicals and energy. For the future, less tangible sectors like service, health care, and finance will get more attention. In the banking sector, the appliance of risk management and monitoring is underdeveloped. Of course, technical instruments can and are copied from Western banking institutions. But more important than possessing the technology is the application of the instruments in commercial situations. The latter is the Achilles heel of Chinese banking industry. Because of the increased intensity of direct foreign investments and cross border mergers and acquisitions, the practical use of these financial instruments becomes relevant for being successful in the coming decades.

The global network of Chinese mentioned before can be seen as a threat or opportunity. On the one hand, the Chinese community prefers to do business between themselves based on their Guanxi. All the acquired knowledge and experience is then transferred to Chinese organizations and networks. On the other hand, the Chinese expats meanwhile have also their interests in country of choice. Very often, these Chinese connections are not properly used by their Western counterparts and opportunities are missed.

Here, the interpretation of time within a society will form a separate megatrend. Time can be regarded to be a production factor and has economic relevance. The time dimension is often neglected when searching for megatrends or creating strategic innovations. The reason is that they are not always so obvious and are often a hidden aspect of economic organization (van Someren 1991, 2005a). For example, reducing waiting time in a manufacturing process can have much greater positive effects than speeding up the process by expensive IT solutions neglecting the waiting time problem. The just-in-time manufacturing system is an illustration of this time-related phenomenon.

Therefore, time-related aspects are crucial for understanding long-term developments and their underlying strategic innovations. In the USA, a 24/7 attitude is part of the way the society is organized. In the EU, the 24/7 mentality is much less present. Holidays in the USA are shorter than in the EU. Holidays in the EU are almost a holy right of the working class. It also explains the difficult political discussions around the topic of retirement age and working hours. But also sustainability leads to a circular time interpretation instead of the Western linear time notion. The background is partly religious. The Christian religion is based on birth–life–death or a linear time line interpretation. In the economy, this birth–life–death scheme has been translated as input–throughput–output industrial mantra. It represents a throwaway society in which reuse or recycling had no place. In Asia, the Buddhist religion has circular time interpretation. Life is a long circle of sometimes many thousands years, and people come back. The harmonious society is based on this central thought of circular time.

2.4.3 Social Trends

The almost unbridled economic growth and hunting for double digit GDP growth resulted in several wanted and unwanted social side effects. From a historical perspective, the regained power pushes China at the top of the nations which other nations respect and sometimes fear. The humiliation of former foreign powers occupying the country and governing large relevant parts of China can now put away as history. The rebuild self-esteem unleashes a feeling of restoring old norms and values placing China back in the center of the world. China regains its status of superpower. As a result, it will increasingly behave like a superpower. Already now successful Chinese entrepreneurs can sometimes become arrogant when dealing with Western counterparts. The paradox here is that adapted forms of Western capitalism are realizing the economic welfare and social well-being of the former communist thought leaders.

The current Chinese people are now focused on raising personal income levels and acquiring personal wealth. Materialism dominates. But Chinese have always been taught to pursue immaterial values as well. The old traditional philosophical pillars of Chinese society such as Confucianism, Taoism, and Buddhism all stress the importance of immaterial alongside the material values. These immaterial values are deeply rooted in Chinese society and people.

Therefore, the economic success now overshadows the immaterial values, but there will be a turning point. The first signs are already coming to the surface. People rebel against local politicians about issues like polluted baby milk and expropriation of land for property projects. It is a rebellion of the have-nots against the haves. The have-nots are aware of their rights and do not accept anymore the overruling of profit-seeking project developers and entrepreneurs. The current gap between the classical haves like politicians, civil servant, scholars, celebrities, overseas successful Chinese, and entrepreneurs and the have-nots will be closed. The have-nots will be more respected. The anonymous have-nots such as the Chinese Gandhis, local heroes, self-made people, and even happy homeless will get face. As soon as you get face, you get power and respect.

The internet and mobile culture will create more space for different thinking and behaviors. These technologies enhance the power of the educated and less educated individuals and of individuals living in urban or rural areas. But the acceptance of a group will stay important in the Chinese society. Network (Guanxi) can be built in a complementary different way, for example, through social networks. The growth of internet and mobile use and the role of Weibo (Chinese Twitter) and Tudou (The Chinese YouTube) will broaden the communication possibilities for Chinese people. But as every technology, it also depends in what direction and for what purpose the appliances of technologies are developing. The current constraints are in the direction of political statements or political movements. But this trend toward a more regulated Internet could be seen on Internet on January 18, 2012 when Wikipedia and other sites were voluntary closed for one day as a protest against upcoming American Internet ruling. Therefore, it is not the technology itself that changes the world but how we apply it and with what aims.

The power of the powerless can paradoxically turn into a determining force for the Chinese ruling leaders. The reason is that not the further economic success will determine the acceptance of the current Chinese governmental organization but the ability to deal with the increasing social inequality and the power of the powerless. The gap between the poor and the rich, the economic and social discrepancy between the Western and eastern part of the country, has already been identified as an important issue by the ruling party.

It is not only the economic and social inequality causing problems. Also within social equal groups, differences come up. The unhappiness of the few and generation conflicts are such social phenomena. The unhappiness comes from an unhealthier longer life, due to pollution and upcoming diseases based on unhealthier food patterns. For example, a higher fat consumption leads to typical Western diseases like high blood pressure and heart diseases.

Furthermore, generation conflicts appear on, for example, caretaking for parents or saving instead of spending money. Family ties are still very strong or even the strongest of all social networks, but at the same time, some profound changes are taking place. In China, it is very common to live with several generations under one roof. The social reason is that children are raised to take care of their parents or even grandparents. The economic reason is that household costs could be reduced because of the spread among the generations. But due to economic prosperity and the upcoming social security system and retirement homes, the elderly are separated from the younger generations. Taking care of parents in a physical form will be replaced by an economic form. The young generation will buy off their duty to take care of their parents. This is a true social revolution.

2.4.4 Technological Trends

Chinese are fond of technology and technical gadgets. For example, the hype of iPhones and iPads has led to small riots around Apple stores. It explains why in the eyes of Chinese, innovation is equal to technology. Technology fits to their preference for products which you can touch and do something concrete. The paradox here is that non-technological part of innovation gets much less attention although the future of Green China will be determined by these nontechnical innovations. Examples are the creation of a better health-care system and service sector. Moreover, the recent policy of indigenous innovation is largely dependent on strategic innovation of which the nontechnical aspects are by far the key success factors.

For Chinese, the practical aspects of innovation are more important than the fundamental inner workings or basic theory. The how is more relevant than the why. This is a defendable position as long as emulation of Western technology is at the core of most value chains. As soon as more radical technologies have to be developed in-house, the fundamental part of R&D instead of improved products becomes more relevant. The creative copycat approach has to be combined with innovative entrepreneurialism.

Powerful nontechnical instruments will be Chinese technology standards and globally organized R&D centers dominated by Chinese organizations. In some areas, these centers have already been set up. The copycat behavior has enabled to catapult the Chinese renewable energy sector to a worldwide leading status. The simple complexity of applying and selling existing technology for windmills and solar panels explains the current success. But the future of the renewable energy sector requires new technologies and business models. The simple complexity will be replaced by complex new strategic innovations.

In case of China, the aim and dream to become an economic superpower is made dependent of becoming a high-tech superpower. Most high-tech areas are announced to be strategic areas such as biotech, high speed, and airplane industry. Space and deep sea technologies are rather underdeveloped industries but will become relevant. On the short term, control of space is necessary for controlling satellites and with that economic and military infrastructure. On the long term, commercial transport through space and even economic activities like labs and mining do have potential. Deep sea technology is necessary for future mining and using natural resources like offshore windmills.

The global policy of China has a multiple aim: becoming an economic and financial super power, a high-tech controller, and Green in China instead of Made in China. Green China will be an active nonmilitary superpower but an active power in regional economic alliances to secure their resources and supplies.

2.4.5 Ecological Trends

Maintaining harmony is one of the oldest principles of living a decent and respectful life in China. Keeping up harmony can be applied upon one's personal life. It explains why Chinese accept a small lie for trying to find harmony. In a Western context, it is more the opposite: Telling the truth is preferred above harmony. But harmony also holds for society at large.

In this respect, sustainability defined as balancing economic, social, and ecological aspects fits very well to China's history and core norms and values. In theory, it fits better to Chinese history than Western history. When looking at the numerous ecological disasters in nowadays China, one may be amazed about the principle of harmony because it seems to be totally absent. That is true; it was in most instances indeed absent in the recent decades. But the notion of harmony does not indicate nor say something about the time aspect. Therefore, it depends on the time horizon when the balance leading harmony should be restored. The decades of economic domination will now be brought more in balance by paying attention to social and ecological issues. The Triple P harmony will be restored on different levels and we give some illustrations.

In different regions in the world, you can find hell and paradise on earth. The same holds for China where very polluted regions and beautiful nature are present. Especially the big cities and the megacities have become more or less point emission areas. Furthermore, many large industrial areas in the countryside have been heavily polluted by all kind of emissions and other detrimental ecological effects. But it would not be China if there are no opposite developments.

China was one of the first countries which has made plans for large-scale eco-cities. On a small scale, many cities around China already took the first step by quickly introducing LED lighting and flexible solar foil on windows of, for example, hotels. These brownfield initiatives were succeeded by the next step of greenfield eco-cities. The master plan of greenfield eco-cities took into account all issues including energy, water, air, mobility, and so on. These plans were soon followed by the next level plans of creating sustainable industrial areas where closed material cycles and interconnected primary and secondary processes between various firms would create an ideal sustainable business park.

Besides city planning, the enormous demand for energy unleashed the creation of companies focusing on renewable energies. Within a decade, these windmill and solar panel companies became world market leader measured in numbers of goods sold. But it was often not the domestic Chinese market but foreign markets which allowed the increase of output. For example, the German feed-in tariff tax system for renewable energy attracted Chinese suppliers. The Chinese overcapacity found a way out in a profitable market.

Besides energy, also water issues like droughts, drinking water, industrial water, and sanitation are of extreme importance for the future of Green China. The water pollution problem is immense and multiplied by the focus on manufacturing activities. With regard to access to water resources, China is in the danger zone which can hamper further economic development. Already since the 1980s, the Chinese government was aware of these megatrend and some measures have been taken. For example, in the Huaihe basin, more than 1,000 paper mills were closed and replaced downstream after a study showed the imbalance between added economic value and ecological damage with a larger negative economic value (Wang 1996). However, these kinds of measures could keep up with the speed, scale, and scope of economic development. But for the future, water management is a core issue.

Of the big sustainable issues like climate change and natural resources, biodiversity is the most difficult one to grasp, at least for business managers. How can biodiversity have an impact on my bottom line is a question on which they have mostly no answer. For food corporations confronted with diminishing stocks of seafood, the biodiversity issue is clear. But for car manufacturers is less clear. Still, the trend of decreasing diversity can have profound impact on agro-food products, and linkages and dependencies between industries based on biodiversity are underestimated. The starvation of bees showed the impact on honey production, reduced food supplies for animals leading to shortages of food and disappearance of natural enemies, and a distorted nature. In economic terms, huge capital losses are the result which can hardly be rebuild when extinction of species is the end game.

The pragmatic Chinese found a market opportunity of man-made nature and man-made biodiversity. Some Chinese institutes started to collect almost extinct species worldwide and by means of a breed program increased the volume of the species. Their motivation was less ecological based but more economic driven. The man-made biodiversity is a Chinese-made trend combining ecological and economic interests.

2.4.6 Political Trends

Before the financial crisis, Western politicians, NGOs, and the general public have made comments about the Chinese political system. Human rights were a favorite topic during official visits. After the financial and euro crisis, the same politicians were begging for Chinese money to solve the euro crisis, and nothing was heard about human rights. On a micro level, during many congresses, one could hear the criticism on the long working hours of employees in Chinese factories. In Western eyes, it was inhuman and should be forbidden. The Chinese interpretation including the factory worker herself or himself is that finally he or she got a chance to leave the countryside and earn money. This start was necessary to build up the manufacturing base.

Table 2.2 Chinese political groups

Color	Group	Agenda
Red	Red princess	Stimulate GDP growth and economic development and making China number one
Black	Black conservatives	Restore old values of communist future of China
Blue	Blue elites	Creating powerhouses in the form of firms and regions
Gray	Gray party	Balancing future opportunities with Chinese governance span of control
Yellow	Yellow progressives	Developing next level strategies and new sustainable growth curves
Pink	Flower power	Fostering new social behavior or fall back in traditional role from a position of wealth

The pendulum is now swinging back a little to restore the balance by demand for higher wages and better working conditions. This is politics in a nutshell. It is also dynamic economic development in a nutshell. It took Europe more than 2,000 years to establish a more or less democratic system, and still, we have to fight wars to secure our oil supplies, and we do business with nondemocratic regimes to get resources or sell our products. Our message here is that one should be very careful with judging others knowing that there is not an ideal state yet. This book is not about judging any kind of political system. We however do need to know what the historical background is, what the current context is, and in which direction it will move and develop.

For this purpose and in order to give insight into the societal background and the different groups, Table 2.2 gives an overview.

Within the Chinese political arena, a wide range of groups coexist with changing influence and power in time. It implies that there is one political party indeed, but there are groups with different opinions and interests about the future of China.

The last decades belonged to the red rights fostering economic development. The red rights are programmatic communists who wanted to bring welfare to Chinese people. Very often, they come from communist elite families which have the privilege for very good Beta studies. They can be found in both central and local governments, and they took their chances after the less successful economic reforms of Mao. Still Mao is respected for his historical achievement of keeping China a united country, giving every Chinese a rice bowl, and his ideal to achieve a dominant position in the world. Only his instruments were not up to date.

The right wing of the party changed that but kept alive the dominant position of the Communist party. There are some reactionary powers who want to restore the old situation and give much less freedom for economic initiatives and development. These are the black conservatives, and they could indeed fall in their own pitfall named the domestic black hole. But they can gain power as soon as the social tensions are not solved.

This puts a heavy responsibility on the current blue elites consisting of governmental and private company leaders to take sustainability seriously. It is their duty and task to transform the Chinese economy into the desired harmonized society where ecological detrimental effects are abandoned and every Chinese can take part in the wealth creation. The rice bowl needs to be replaced by feast meal. And everybody must have the opportunity to sit at the table. If they fail, China will fail.

The blue elites can only succeed if the gray party participates. Only a few of the gray party are members of the blue elites; the others are middle class or belong to the lowest level. But the gray parties still have, based on their seniority, a relatively substantial influence on decision making in their families and networks. Based in numbers, the gray ones have little formal high position, but they are the biggest group. It is the mass and their relatively low income being a problem for current officials that makes the power. They are also respected, but the future gray ones are the ones who have built up nowadays China.

The yellow progressives being the civil servants and publicly owned company managers will carefully follow these shifts in power and direction. The yellow progressives are responsible for the creation and development of future laws, industries, and foreign activities. Sustainability will be key in their plans, and even more importantly, the aim of innovative society requires certain freedom of the mind. Only then the new industries will foster and be a rival to Western industries. It is again balance act of central power versus the power of individual creativity resulting in a prosperous sustainable shining China or a clouded polluted China. In this respect, it might be the flower power which can form the linking pin between all the other color groups.

Women represent the pink flower power, and they show two opposite behaviors. On the one hand, women now take their chance and are the driving force of becoming independent of traditional family hierarchy and domination of the men. They become active in government, government-owned organizations, and private business and spend their money. On the other hand, women also fall back in traditional role of staying at home and taking care of family. It also implies that in the background, the women are the decision makers or at least take big influence on relevant future issues and not only the men. Flower power grows everywhere, and the flowers are stronger than in the past. Therefore, feminization in China takes another road than in the West.

2.5 The Challenge of a Colorful China

The greening of China is not a stand-alone topic in which sustainability can be separated from other renewal processes but a solution for the colorful future of China. But even if we understand and know all the trends, it is not a ticket to success. The shifting international and national political colors force us to operate in ever changing conditions. It is like the metaphor of the rainbow colors expresses: The future can be peaceful and harmonious as the Chinese advocate in their official statements. But the opposite is still possible as well, and Chinese leaders know this very well but do not express it. The creation of a sustainable China is the rainbow treasure, but we know we have to work and fight for our treasure.

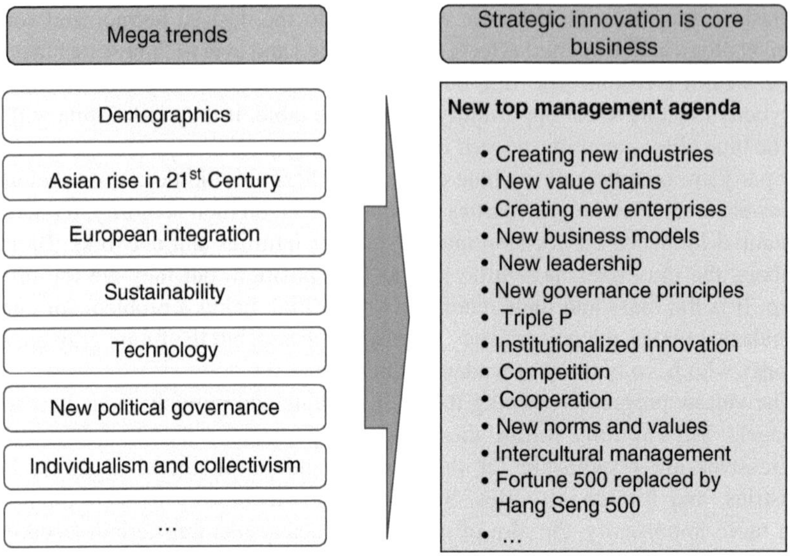

Fig. 2.4 Key challenges for next 20 years. (Source: van Someren, 2010)

The rise of China will be green indeed, but the way sustainability will be realized has connections with a broad range of other topics which together form the setting of the start toward a future world order. These aspects should be integrated in our strategic answer; otherwise, we try to integrate sustainability in our Western way of thinking whereas a broader perspective is necessary. Only then we are able to create future market structures, organizations, and behavior which will lead to the desired outcome of a sustainable world. The wider interpretation of sustainability leads to the integration of following topics which will determine the future world order and which need all an answer. In Fig. 2.4, these are mentioned.

In short, the relevance of each topic can be made clear in a few statements. Demographics determine societal development due to age structure, ethnic and cultural features, knowledge level, labor supply, future demand for products and services, and its accompanying resource intensity. Asian dominance respectively Chinese supremacy is very likely, and this will change how the world will tick. European integration especially on a political and economic governance level will be key for European key member states.

With the rise of Asia but maybe also Arabian, South American, or even African states, Western norms and values will become less prominent than in past centuries. The same holds for governance structure. The increase of the participation of other than Western countries in world economy requires systematic innovation otherwise no position can be conquered, maintained, and enlarged. The core sustainability issues as mentioned in Fig. 2.4 are connected to these other challenges. All these challenges can only be realized when strategic innovations are introduced. Strategic innovations are necessary in every society and private as well as public

Table 2.3 Green China challenges based top management issues

Green China challenges	Top management issue
Government	Adapting to changing laws and regulations
	Governments as partners
Market development	Awareness of Chinese for nontechnical solutions
	Coping with protected markets and supported Chinese firms
Technology	Chinese requirements and specifications
	Focus on non-imitable innovations instead of technology
Finance	Access to Chinese finance sources
	Avoiding black hole and twin star scenarios
Business model	Profitable business in China is key question
	Create Chinese business model
Cross-cultural management	Leading the Chinese
	Dealing with Chinese business partners

organization. These challenges will not be the hypes of the future but key areas for future prosperity.

In this context, Green China is not a separate or stand-alone topic but is an integral part of all the necessary changes toward future of the West and China. This holistic way of thinking fits with Asian way of dealing with larger issues. But Westerners will have to get used to this integral approach and abandon their disintegrated way of solving problems. Sustainability is not only about emissions and CO_2 but about dealing with worldwide key challenges related to new world order of doing business.

But sustainability is only one top issue; others are China strategy, creating global corporations and institutionalized innovation. With the theme of Green China, these topics come together simultaneously. These key challenges change top management agenda. This change of perspective is crucial in finding the right answers.[2] In Chinese words, this rainbow is a simplification of the ultimate dream of a harmonious society. Harmonious society within the rainbow implies a trade-off between different interests and performance indicators. This is exactly what a green society is about because sustainability boils down to the trade-off between and mutual reinforcement of social, ecological, and economic dimensions.

The challenges of the Green China require a specific top management agenda as mentioned in the previous section. In Table 2.3, a few examples are given.

Taken together, the Green China challenges form a separate universe with their own rules.

The Green China challenges confront Western organizations with unique top management issues. The Chinese government will of course focus on Chinese interests. But Chinese governments are not always on one line with each other, and governmental organizations can be rivals of each other. Moreover, laws and

[2] van Someren, 2005a.

regulations can be used against foreign intruders. Still you need government as your partner. The access to and development of Chinese market will partly be based on technology. The better Western organizations are able to mingle technology with nontechnical solutions, the better the chances of survival. This combination will also be a force weapon against governmental supported Chinese firms. The ability to adapt to Chinese requirements and specific demand combined with less easy to imitate innovations keeps Western organizations in the race with Green Dragons. Only organizations who know how to play the Chinese game in cooperation with Chinese business partners will have access to Chinese finance sources. The greatest risk for Western organizations is to fall in the trap of black hole and twin star scenarios. Above aspects require a Chinese business model, but for business, there is only one rule: Earn money and make profit. Cross-cultural management is necessary to understand Green China and to get a part of the rainbow treasure.

Therefore, the Greening of China implies revolutionary changes on many fronts which will dramatically change our world. Green China is about roles of government, alternative ways of market organization, and roles of networks, individual organizations, and individuals on a global scale. In fact, it is about a new world order. The West and China will have different views on this new world order.

References

Kleining, J. (2008). Dispersed economic power? Overseas Chinese between discrimination and success in business. Konrad Adenauer Stiftung, Berlin

Naisbitt, J., & Naisbitt, D. (2010). *China's megatrends: The 8 pillars of a new society*. New York: Harper Business.

Hefele, P. & Bade (2011). *Re-inventing the city? Domestic migration and urbanisation in the People's Republic of China*. Berlin: Konrad-Adenauer-Stiftung e.V.

Someren, T. C. R. van (1991, June). *Innovatie, Emulatie en Tijd. De rol van de organisatorische vernieuwingen in het economische proces*. Tinbergen Institute Research Series, no. 9, (diss.).

Someren, T. C. R. van, & Nijhof, A. (2010). Triple P business development in the Dutch agro-food sector. Van Gorcum Publishers. Assen.

Someren, T. C. R. van, & Someren-Wang, S. van (2009, September 22). Nederland moet zich spiegelen aan China. *Financieele Dagblad*, p. 6.

Someren, T. C. R. van, & Someren-Wang, S. van (2011, December 2). Kans na harde landing. *Financieele Dagblad*, p. 7.

The China Greentech Report (2009). MangoStrategy.

Someren, T.C. R. van (2005a, May). *Strategische Innovationen. So machen Sie Ihr Unternehmen einzigartig*. Wiesbaden: Gabler Verlag.

Wang, S. (1996). *Study on Chinese water environment carrying capacity*, Beijing, China: Beijing Normal University.

The Threats and Opportunities of Green China

3

See Through the Five-Year Plans for a Sustainable Future of the USA and the EU

In the West, the sheer size of China leads to predominantly two kind of reactions. China is a threat, or China is an opportunity. Both interpretations are not right. Seeing China as a threat implies underestimating your own possibilities and overestimating the power and capabilities of China. Seeing China as an opportunity is naive and underestimates the potential rivalry and cutthroat competition of existing and future players. However, what is true is that China is a big challenge for the West. The challenge is on five fronts: mutuality, money, markets, manpower, and mentality. China has a tradition in coping with these issues. The famous five-year plan and many other plans are the instrument of central government to put forward the ambitions of China. The core of the 11th and 12th five-year plans is on creating a harmonious society in which sustainability has a prominent place. But sustainability is translated into economic plans for growth and development linked with large investments in the industries of the future. The West is challenged in the heart of their business: innovation. Green China is about creating new industries and firms of the future. The West showed signs of a dying knight. In the legend, the only way to sustain the life of the knight is to bathe in the blood of a dragon. But can the West find the blood of a Green Dragon?

3.1 From Sleepy Dragon to Saving Angel? The Attraction of China in History

"The Chinese economy has been industrialized, commercialized, monetized and urbanized far ahead of other countries. The manifold signs include among others are merchants, entrepreneurship, the creation of private economic organizations, the increased complexity of structure of business, growing density of the local market network, new structures to handle large numbers of employees, an international division of labor, absolute and relative productivity in manufactures and

T.C.R. van Someren and S. van Someren-Wang, *Green China*,
Management for Professionals, DOI 10.1007/978-3-642-28810-4_3,

productive superiority. Furthermore, technological innovations were needed to fuel the increased output and export. Imitation and in the end by improvement these techniques contributed to the excellence of products."[1] You might think I read nothing new here; it is a usual description of the Chinese economy I can read almost every day on Internet, Twitter, or in the newspaper. But did you know that this quote from scholar Mark Elvin described the Chinese economy in the twelfth till fifteenth century? And at that time, it was not the Chinese who copied techniques, but it was the Europeans who stimulated their progress by copying and improving Chinese technology. China among other Asian countries such as India had a strong position in the world economy for many centuries till about the sixteenth century or even to some historians till 1750.

The "Asiatic mode of production in China (AMP)" enabled Asian economies to grow faster and more than Europe (Frank 1998). After this unmatched long period of China ruling the economic waves, they were surpassed by many other countries. The Portuguese first took over the reign of trade of the Spaniards. Soon, Portugal was on their turn replaced by the Dutch. The Dutch world empire lasted not even a century till the UK took over. But also the Britons were passed by the USA who are in our time measured by GDP or Fortune 500 companies, the superpower. Also on the level of individual firms, this life cycle of up- and downturn occurs. For example, the Portuguese individual traders were surpassed by a new multinational form named the Dutch East Indian Trading Company ("VOC"). This first corporation of the world wiped away the Portuguese. Nowadays, the British supermarket Tesco took over from long-time market leader Sainsbury. This pattern of rising, growth, slowdown, and fall of societies and firms is like a team pursuit. Each time, another country or industry leader is in the lead.

But measured in the time scale of the Chinese society, which exists over 5,000 years, was this relatively short period of a few centuries only a short intermezzo of the West? Will the Chinese regain their lost position? Would China be the first country to recapture the position of the leading economy or offer the current global economy opportunities for everybody? Do the Western companies have a chance in the dynamic Chinese economy?

The position of the West is nowadays much better than about three to four centuries ago. These days, in many sectors, the West still have, for example, better technologies and better distribution systems, and they have in almost industry a large bunch of small and large multinationals with superior technologies embodied in a large variety of products and services. A few centuries ago, Western traders in Asian markets had nothing to offer except the silver extracted in the colonized Americas. Nowadays, trading is in fact nothing else than swapping market for technology. Chinese allow Western companies to enter their markets in exchange of technology. But is it technology alone that will the West give a chance to play a substantial role in the future world economy? Swapping technology for markets is a

[1] Frank 1998, quoting Elvin (1973) p. 218–219.

short-term game and a very dangerous game. Nothing else can be copied as easy and quickly as technology. Therefore, the central question is as follows: Will the current economic dominance of the West indeed be an intermezzo or can Western companies create new opportunities in China? The answer is depending on a few topics that will be treated in this book. At least a more profound knowledge of the inner workings of the Chinese economy and the Chinese business mind is needed. Chinese are masters of long-term thinking and achieving long-term goals. Of course, you knew. Chinese are also masters of gaining short-term gains and flexible solutions for new unexpected situations. For Westerners, this kind of paradox behavior of Chinese is difficult to grasp. Therefore, for Westerners doing business in China, it is absolute necessary to know what China's leaders want to achieve and what they have in mind. Only then individual firm's China strategies get meaningful and can build on a win–win basis. Win–win is in the West a beloved notion. But win–win for whom? And for how long? And at what price? And do Chinese have the same interpretation of win–win? Pursuing your own profit or sales targets using your existent portfolio of products and services will not be enough for a long successful stay in China. Understanding the development process and the accompanying governmental policies is crucial. The move of China to another growth curve will have a profound impact on the world and in fact creates a new world order.

3.2 China: A Threat or an Opportunity?

3.2.1 China as a Threat

In the West, the rise of China mostly elicits two reactions. The first is one of fear about a potential superpower. Inward-looking countries with little knowledge and empathy about the outside experience a threat. The second is one of excessive admiration. When top managers and politicians visit China, they cannot hide their admiration for the achievements and speed of development. But when asked how they will change their own business or country in a similar way, you often only hear the wind through the trees.

The sheer size of China for most Westerners, especially those from smaller countries, makes a deep impression. Examples and comparisons are used to make a mind-boggling picture of the dimension of China. The EU and China are trading more than one billion euro a day. About half of China's exports are currently produced by foreign-invested companies. Other Asian companies from Japan, Taiwan, Hong Kong, and South Korea play a great role in this foreign direct investment (FDI). Only 1.7 % of the total flow of FDI into Europe comes from China. The biggest company of the world measured as market capitalization is PetroChina with $343.3 billion (2009). The second biggest employer is State Grid from China with 1.5 million employees after Walmart Stores with 2.1 million employees. And so on and so on. But first, the Green Dragon is not unbeatable, and second, the Green Dragon is not without failure.

Table 3.1 Perceived threats and opportunities of China

China as a threat	China as an opportunity
Yellow danger due to its population size	Cheap labor provider
Military force	Potential markets
New geopolitical center	Test-market new technologies
Copy–paste mentality	Design paradise
Economic power	Size and scope of R&D potential
Deep capital pockets	Greenfield opportunities
Environmental pollution	Investment gold mine

Despite the opening, fast growth of China, and all the Made in China products in our hands, the Western thinking about China is in its infancy. When we are honest, our thinking is still black and white. There are two groups. The one group thinks China is on big threat, and the other group only sees opportunities. In Table 3.1, some common illustrations are presented.

First, we have a look at China as a threat. When China awakens, the yellow danger will conquer the world because their size of the population dwarfs most other nations. However, due to the one-child policy, the absolute number will decrease, and in the coming decades, also the aging problem will increase. Like in Europe, the question arises who will take care of the elderly people, and the number of fee earners against the number of retired people will decrease. Moreover, India will take over as most populated nation. Even in the West, these problems of aging are not solved yet, and opportunities of learning from each other exist.

China as a military force will indeed be built up. But when looking at the history of China, military force has been used for internal fights and fighting foreign armies but not for conquering other regions by force. Of course, China's interests in our current global world are increasing rapidly, and it remains to be seen whether military force will be used to defend their interests. Besides global interests, China and their neighbors have the opportunity to develop toward a geopolitical superpower. But it must not be forgotten that all the countries in Southeast Asia are not a unity. On the contrary, they are divided, and some of them even fear the power of China.

Above all, the Southeast Asian countries have no serious attempt to start a unification project like the EU is trying for years. The ASEAN countries consisting of Brunei Darussalam, Cambodia, Indonesia, Laos, Malaysia, Myanmar, Philippines, Singapore, Thailand, and Vietnam do have programs to integrate their regions and to take away trade barriers. Simultaneously, border conflicts between Thailand and Cambodia exist.

A true problem is the copy–paste mentality. But imitation is in the core of Chinese behavior and cannot be turned off easily and immediately. But one should not forget that imitation is part of us all starting at home by learning from parents and in school from teachers. Already China itself has a problem with imitation in two ways. First, imitation results in overcapacity and losses. Imitation leads to the famous pork cycle of high prices followed by low prices. High prices and high profit margins attract others leading to a low-price high-volume market. Second, as soon as China develops into a knowledge economy driven by innovation, imitation will become the enemy of the entrepreneur.

One should more fear the Chinese economic power than the military power. Already now, the economic power based on low cost and high volumes fueled the trade statistics between Europe and China. On top of that, the earned money with exports has filled the Chinese Treasury with large sums of capital. These deep pockets with hard-earned money can be used to exercise economic power play. But of course, the fast growth and development has led to ecological detrimental effects. About 70 % of the rivers in China are heavily polluted; in most cities, air pollution is a big problem, and access to fresh (drinking) water causes problems for industrial development as well as human consumption. Not only the factory of the world is threatened but also the whole world is influenced by the Chinese pollution and use of resources. These detrimental effects can lower GDP growth by 2–3 % which makes immediate action necessary.

3.2.2 China as an Opportunity

Like China as a big threat, the other side, China as one big opportunity should also be judged in balance. For the past decades, China has been the cheap labor provider of the world. But labor prices rise, and the workshop for the West will move from Made in China toward Designed in China.

Moreover, education levels will rise, and cheap labor will only be remembered as the start of a growth cycle and not the end of it. With rising spendable income, every Western businessman's dream is the same: Selling to Chinese makes me rich. Bottom line, despite all PowerPoint slides and business plans with difficulties, bottlenecks and risks, and if they are honest about their real feelings, all single entrepreneurs, SMEs, and top managers of MNEs think like that. Of course, every entrepreneur should be optimistic, but in case of China, it is good to be a realist as well.

In most cases, there are no golden mountains coming toward you or to be mined easily. In Europe, it is hard to discover any golden mountain. In many cases, it is difficult to create a golden mountain because due to the high density on any kind of regulation, it can be made very difficult to implement new technologies or products. Some companies now even start to use China as a launching market for new technologies or concepts instead of their own home market. The hope is that a launch in China opens up possibilities in Western markets.

For some, China might become the design paradise because the Chinese are fond of new architecture and they like new gadgets. It seems like everything is possible and the future is there. For some, this might be the case; for the most, it will be difficult to make profit because not all Western ideas will be accepted and some will be copied by Chinese designers.

In the same line of thinking, the potential for technical educated people is huge. Each year in China, about 1.5 million technicians finish their studies. Additionally, many students are attracted by Western universities to study or finish their PhD.

For Western universities, these cohorts of students are interesting because they earn money from every single student. Increasingly, R&D laboratories with relatively cheap researchers are founded. So there is not only a design paradise but also a R&D paradise. But it will only become a paradise when the road to hell can be prevented. The road to hell is when the researchers are educated and enabled to make mistakes in the first years of their career before returning to their Chinese home country.

The same holds for greenfield operations. Here are lot of opportunities to start greenfield operations greatly supported by business partners and governments. But there is always a Chinese price in the form of limited growth or copycat behavior opposite by the brother or nephew of your alliance partner. And it will be no surprise anymore; there are a lot of investment opportunities in every corner of the country and in most industries. But it depends on reliable partners and information about opportunities, and the quickly changing society and its rules make it difficult to sustain your profits.

3.2.3 Heaven or Hell?

So what should we now think of China, based on the threats and the opportunities, will it be heaven or hell? Both the pessimistic, China as a threat, and the optimistic variant, China as an opportunity, are the wrong perspectives to deal with China. Western thinking about threats and opportunities is biased by our binary way of thinking instead of finding the balance or dealing with paradoxes. In Fig. 3.1, this is demonstrated with help of the growth curves.

The view of China predominantly as threat exaggerates the power of China and underestimates its own countervailing possibilities. The other way round, seeing only the opportunities makes one blind for risks and capabilities of Chinese parties to take away business from you. Both options ensure your existence in the short term but not in the long term. Only a balanced view of threats and opportunities of Chinese rise will enable you to increase your chances in a right decision. Only then, it is possible to see which projects, markets, products, or investments will support a new growth cycle. There are no golden mountains only for you, but Green China offers everyone to make good business. But balancing the threats and opportunities and still making a profit requires constantly adaptation to changing circumstances and in most cases strategic innovations.

3.3 Why Do China and the West Need Each Other?

The balancing act of threats and opportunities and transforming them into challenges with prospects of value conversion depends on 5 Ms: mutuality, money, markets, manpower, and mentality. These 5 Ms determine the conditions for strategic innovations and value conversion ability of Green China. Hence, the 5 Ms are not only relevant for China but also determine the conditions for the EU and

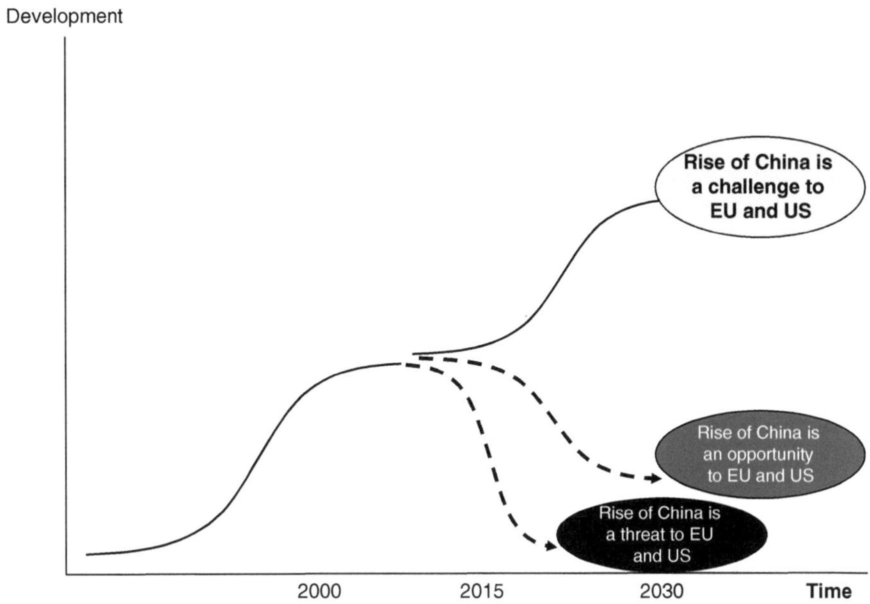

Fig. 3.1 Threats and opportunities in relation to the growth cycle. (Source: based upon Van someren, 2005, Strategische Innovationen, Gabler Verlag)

the USA to participate in the prospects of Green China. The 5 Ms will be elaborated shortly in the next sections.

3.3.1 Mutuality: The Interdependence of Western and Chinese Business

In 2009, the USA and not China was still the biggest manufacturer of the world. The manufacturing output of the USA was $1.682 billion and of China $1.481 billion.

About 30 years ago, a bipolar world between the USA and Europe existed. With the rise of Japan and their export machine, soon a tripolar economic world has been shaped. China played no role yet, but around 2010, a multipolar world is in the making because the trade volumes with BRIC countries and especially China have exploded. In Table 3.2, the five main import and export partners of the EU27 are summed up.

Table 3.2 demonstrates the dominant position of China with nearly 20 % of both import and export volumes besides the USA. When extrapolating the growth of these trade volumes, China will emulate the USA very soon on both trade streams.

Table 3.3 shows a similar pattern with regard to the position of the USA and China.

Table 3.2 Major import and export partners of the EU27

The major import partners (2010)			
Rank	Partners	Mio euro	%
	EU27	1,501,844	100
1	**China**	**282,011**	**18.8**
2	USA	169,467	11.3
3	Russia	158,385	10.5
4	Switzerland	84,126	5.6
5	Norway	79,179	5.3
The major export partners (2010)			
Rank	Partners	Mio euro	%
	EU27	1,348,792	100
1	USA	242,095	17.9
2	**China**	**113,118**	**8.4**
3	Switzerland	105,433	7.8
4	Russia	86,509	6.4
5	Turkey	61,190	4.5

Source: European Commission

The import and export data lead to the conclusion that at the moment of writing, there exists a mutual dependency between both the EU and the USA and China. However, from a dynamic point of view, as the strategic innovation theory advocates, the volume and nature of this mutual dependency will change in future. The reason is the rise of Green China. The current major imports are largely based on low-margin products like shoes, toys, and of course high-margin products like reimported mobile phones, flat screens, and so on. The export consists of all kind of products like agro-food, cars, and luxury products.

Because of the rise of Green China, the import–export trade will partly shift to environmental technology and increasingly services. Depending on China's development, products from emerging industries like greentech, nanotech, biotech, ICT, and batteries might complement existing trade. Another aspect is the transfer of knowledge and experiences which are transferred from European firms to Chinese firms as part of cooperation agreements. The value of this value transfer is not captured by the statistics but through the accumulation effect that will have a profound influence on current mutual dependency.

But there is something which makes the statistics remarkable. The export volume of the EU27 to Switzerland is nearly as high as to China in 2010. In other words, the potential to increase trade with China has a large potential for growth. The Europeans should carefully look at the strategy of Han Feizi who recommended "to use somebody's spear to attack his shield" like the Chinese do with their own trade. But Europeans could also copy Isaac Newton who said, "If I have seen further, it is by standing on the shoulders of giants."

In this respect, the division of the total trade volumes in goods, services, and foreign direct investments enables a more detailed analysis. In Table 3.4, the trade

Table 3.3 Major import and export partners of the USA

The major US import partners (year to date Nov 2011)			
Rank	Partners	$ billion	%
	All countries	2,023.6	100
1	China	366.5	18.1
2	Canada	290.0	14.3
3	Mexico	241.8	12.0
4	Japan	116.9	5.8
5	Germany	89.5	4.4
The major US export partners (year to date Nov 2011)			
Rank	Partners	$ billion	%
	All countries	1,354.1	100
1	Canada	258.3	19.1
2	Mexico	181.2	13.4
3	China	94.2	7.0
4	Japan	60.8	4.5
5	UK	51.1	3.8

Source: US Census Bureau

Table 3.4 Trade in goods, services, and foreign direct investment between the EU and China

2010 in billion euro	Trade in goods	Trade in services	Foreign direct investment
EU exports to China	113.1		
EU imports from China	281.9		
EU exports to China		20.2	
EU imports from China		16.3	
EU inward investment to China			4.9
China inward investment to the EU			0.9

Source: European Commission

volumes and FDI between the EU and China are presented. Table 3.5 shows the FDI between the USA and China.

As can be seen from Tables 3.4 and 3.5, the potential for services and foreign direct investment is high and untapped. Especially the discussion or better fear in recent years about the threat of Chinese taking over by direct investment is not justified. The investments of European organizations in China are more than five times higher than Chinese foreign investments in the EU27. But Green China will change these statistics. But it depends on the nature and level of the necessary strategic innovations in both the EU and China in which direction and to what extent the trade in goods, services, and foreign direct investments will change. Moreover, the Chinese government has large sums of capital available, and they have several options how to spend and invest these enormous capital reserves. The next section will discuss the options, and the reader should keep in mind the numbers given in Tables 3.4 and 3.5.

Table 3.5 Bilateral FDI of the USA and China and cumulative value at end of 2009

In $ millions	2003	2004	2005	2006	2007	2008	2009	Cumulative value of FDI at end of 2009
China's FDI in the USA (historical cost basis)	−62	150	146	315	137	368	−271	791
US FDI in China (historical cost basis)	1.273	4.499	1.955	4.226	5.331	15.726	−6.997	49.403

Source: US Congress Research US Bureau of Economic Analysis

3.3.2 Money: Saved by the Poor, Gambled by the Rich, Allocated by the Government

Irrespective whether you call it welfare or prosperity or growth and development, in the end, it is all about money. The earth turns around the sun, and Green China turns around money. Did we not hear about Triple P and the ecological aspects and social values and happiness? Of course, we did, but new growth cycles start and develop with ideas and money which in a sustainable way can lead to happiness.

Green China is about elevating a nation not only above the poverty limits and status of a developing country. It is also about the capacity of value conversion and creating new sustainable industries and building a sustainable society. In this respect, the Chinese government has chosen for a specific approach. The first step of imitation was to open the country and to be able to earn money with attracting manufacturing activities which required cheap labor conditions. The second step of improvement was to learn from the incoming foreign organizations. The third step of innovation is to move on to the stage of becoming an innovative nation followed by the fourth step of internationalization. For China as a whole country, this so-called 4I development path is now more or less between the second and third stages. Some firms or even subindustries are more front-runner or laggards.

In the first two phases in the past decades, the Chinese economy has generated and accumulated large sums of capital in the hands of both the government and some entrepreneurs. Also, the EU has had prosperous times in which new wealth has been created. On both sides, the majority of the people worked hard, and some got more than the others. But the financial crisis, debt crisis, and euro create huge uncertainties in all markets, especially in the financial sector. Creating a sustainable world cannot be achieved based on new principles like Triple P and care for nature alone. The new revolutionary growth curve of sustainability as shown in Chap. 1 implies new investment opportunities. But the years as of 2008 are linked with crisis and not sustainability. The good news is that these extreme and contradictionary situations are a normal part of the macro changes. But the bad news is that some gain and some lose in this period of change.

In fact what happens is that the accumulated capital has to find ways for investment which should go beyond supporting banks by governments. The accumulated money and capital are earned by entrepreneurs shaping new businesses and realized by employees, saved by the poor, gambled by the rich, and allocated by the government. For example, the EU estimated that it needs more than 300 billion euros only for building a smart grid for transporting renewable energy electricity streams.

The Chinese government has accumulated more than 3.200 billion dollars which need investment opportunities. With regard to the rise of Green China, there are five strategic options on how to deal with this mountain of money.

The first is to invest in worldwide markets for bonds, currencies, gold, and oil. But the large state funds containing more than 3.200 billion dollars are too big to invest in this first category without disturbing those markets. Therefore, China will buy bonds from European member states for strategic reasons. Especially, in conformity with Sun Tze, the weak countries like Greece and Portugal will be targeted. The weak underbelly forms a first foothold. In this category, the indirect route is to fund other countries through the IMF and gain influence.

The second is domestic investment in infrastructure, social security, pension funds, and health care. This is a real option and already taking off, but also here, too big inland investments destroy domestic economy as the housing market bubble and overcapacity in many industries already show. The high risk of inflation can destroy the value of the reserves.

The third option is international investments. The only markets left are natural resources, public procurement, and corporate investments. Capturing natural resources ensures the lifeline of domestic manufacturing, and mergers and acquisitions ensure technology and future entrance into Western markets. The majority share of these funds, around 75 %, has been invested in the Asiatic region. Now the EU, the USA, and other geographical areas will be targeted. Meanwhile, Chinese firms supported by government openly compete for public procurement contracts, for example, building roads.

The fourth option is not to invest in the supply side of domestic industries but in building up innovation capacity. The creation of national innovation requires lot of investments in R&D center and educational programs.

The fifth option is that the money of state funds is needed for backing up banks with bad loans. Not only the government stimulated Chinese banks to lend money to invest in all kind of projects but also private lending and borrowing between network members prospered. But the expectation is that a substantial part of the loans will have to be written off due to loss projects.

The policy appreciation of the RMB helps the third global strategy of capturing natural and knowledge resources and launching China as the world's dominant economy. Here, a strategic shift is possible where micro, meso, and macro measures reinforce each other and the Chinese central government operates as the great governor of markets. The current success of China's growth was based on low-cost, high-volume, low-profit margin manufacturing for export supported by a low RMB. A rapid increase of the value of the RMB would destroy low-margin export industries. The future of China's growth will be based on high-value, high-innovation, corporate market supported by a higher value of the RMB.

The current Chinese production mode will lose its immense competitive power due to rising labor costs. In 1990, China's share in world output was around 3 %, and in 2011, it is near to 20 % and slightly higher than the 19 % of the USA. Currently, Chinese wages are rising with 17 % annually, and in Shanghai, wages rose by 300 % from 2003 to 2010. Chinese average wages are on the level of Mexico. In the USA, wages rise by 3 %. Even for higher management, Chinese wages are rising faster than Western levels. Moreover, productivity levels in the West are higher than in China. Together, the USA and Germany take care of 80 % of global manufacturing value.

Furthermore, the higher value of the RMB makes global mergers and acquisitions cheaper and enables China to grasp high-end markets. But buying technology will not be sufficient; also the ability to innovate within Chinese individual firms, clusters, and networks will be necessary. Particularly nontechnical innovations like building an innovative corporate culture, organizational innovations, Triple P performance measures, logistics, branding, and cross-cultural management will prove to be crucial. Future strategic innovations applied by firms and policies will contain a mix of these micro, meso, and macro levels. It is necessary to understand these different approaches which all will determine the future of business world.

As the coming chapters will reveal, on all levels, different approaches and choices with regard to context and strategic innovations are applied in China and the West. But there is one golden rule which applies in all times and on all levels: Every action must eventually lead to value conversion.[2] Value conversion originally focused on profits, but it can also comprise social and ecological aspects nowadays known as Triple P (people, profit, and planet). Strategic innovation in business should lead to value conversion, but also nation states require an earning capacity as a whole in order to contribute positively to society and the world.

3.3.3 Markets: The Declining Market Versus the Chinese Purchase Power

Green China requires many strategic innovations in every industry like renewables, construction, transport, and banking. Sometimes, complete new (sub)industries and firms will be founded such as with nano- and biotechnology. In China, huge new market opportunities arise, but due to the Chinese mentality, many entrepreneurs will take their chances. Hence, for Western organizations, it will be extremely difficult to get a piece of the growing cake and new cakes coming from the Chinese oven.

It is not only the expectation of cutthroat competition but also direct or indirect protection of markets by Chinese government. For example, in the first wave of

[2] H.W. de Jong 1985, T.C.R. 1991, 2005.

solar panel manufacturers, many newborn producers have difficulty with getting profitable business. In their tradition, government will take action and take measures like forced mergers or conquering foreign markets with governmental support. So again, the threats and opportunities have to be calculated precisely, and flexible strategic plans are needed to survive in Green China.

For the Chinese government, contrary to old industries like steel- and fossil fuel-based car manufacturing, the new industries such as renewable energy, electrical cars, nano- and biotech, agro-food, and water are extremely relevant for Green China's future. The reason behind is that the old industries cannot be emulated anymore, and in some cases, it is even not very wise to put your money in dying subindustries. It is better to invest and educate your people in electrical cars and battery technology than in gasoline technology. Therefore, the Chinese government is willing to foster and protect these promising industries of the future in order to get an advantage on the West and to emulate their nations and companies.

3.3.4 Manpower: Quantity and Quality

The size of the population itself says nothing about the ability of creating innovations or value conversion possibilities and earning capacity. The absolute number can be an indication of power. But China has demonstrated itself the difference between before and after opening markets. The achievements of the same population in different contexts are huge. Therefore, the core issue of manpower is less quantity but more quality. It is not only absolute numbers but also their qualifications and their occupation. What do you do with your knowledge and your available time? Go on holiday or create a new company? Moreover, even without knowledge, the Chinese were able to earn lots of money. So there is more than nowadays popular mantra of knowledge as the most important production factor of the twenty-first century. In the next section, we will go into the background; for now, we stick to the aspects of knowledge and education.

With regard to knowledge, even a highly educated country is not an assurance for growth and development. It depends on what education has been followed and in what and where the highly educated people apply their knowledge. The so-called L to E ratio illustrates this aspect of the quality of manpower. The L to E ratio is the ration between the number of lawyers and the number of engineers occupied in a nation. It might not be precisely right, but it shows the priority in direction. The ratio of engineers to lawyers graduated each year in Japan is 10 to 1 whereas the ratio of engineers to lawyers graduated each year in the USA is 1 to 10. The assumption is that lawyers cannot and will not create new products and a society full of lawyers will lose the race in becoming more competitive. In the USA, famous for its claim culture, its culture of very detailed written contracts and suing each other, the popularity of lawyers is not a surprise anymore. It is feared that the claim culture will swap over to Europe and studying any kind of technical study is in many countries not popular. The L to E ratio would increase further.

Table 3.6 Leaders of standing committee and their educational background

Name of leader	Main educational background
Hu Jintao	Tsinghua University, engineer
Wu Bangguo	Tsinghua University, engineer
Wen Jiabao	Beijing Institute of Geology, engineer
Jia Qinglin	Department of Electric Power of Hebei Engineering College, senior engineer
Li Changchun	Department of Electric Machinery of Harbin Institute of Technology, engineer
Xi Jinping	Tsinghua University, doctor of laws (LLD)
Li Keqiang	Peking University, doctor of economics
He Guoqiang	Beijing Institute of Chemical Engineering, senior engineer
Zhou Yongkang	Beijing Petroleum Institute, senior engineer

There is also a link between innovation and the L to E ratio. Worldwide, the number of patents is increasing due to R&D, increasing number of countries relevant for protecting your knowledge and the smaller inventions and improvements laid down in patents. In the USA, some already talk about the IP bubble. When we apply the "L to E ratio" to the increasing number of patents and the IP bubble, it can be expected that the L to E ratio will increase. On the long term, the L to E ratio can have some serious impact on the innovation output. With rise of Green China, this trend is not a very good prospect.

What about China? In China, learning and acquiring knowledge is since Confucius a tradition. Nowadays, especially technical studies are very popular because they enable you to create something concrete and practical which you cannot with law study or psychology. Building roads, bridges, houses, cars, and refrigerators is thought to be relevant for building up a society and giving the people work and income. Therefore, the majority of Chinese leaders have a technical background as Table 3.6 shows. Table 3.6 contains the main educational background of some leaders of the standing committee.

The technical studies dominate. Most other leaders in government and business who had the opportunity and luck to study mostly preferred a technical study. This is a contrast to most Western governmental and business leaders. In many European countries, alpha and gamma studies like political science, law, economics, and history are far more popular than the beta studies like chemistry or material science. Also the background of many top managers in private industries follows this pattern. Of course, there are exceptions like in the German car or chemical companies where beta studies still dominate. Is this a conclusion that the Europeans should be studying more beta studies? Yes, for a part, that is an old conclusion drawn by many other politicians and business leaders. But this is only half the story. The other part is that future Green China will increasingly need more leaders with knowledge about, for example, branding, marketing, management of nontechnical innovation, and cross-cultural management. May be it should not be the L to E ratio which should be increased. The Chinese might have a need to develop a nontechnical innovation to engineers ratio, a NTI to E ratio.

3.3.5 Mentality

In the previous sections, the issues of manpower, education, knowledge, and money were shortly discussed. But all these ingredients remain dead capital as long as creative entrepreneurs are absent. Only innovative entrepreneurs are able to combine these resources and in time new products, services, and business models. Entrepreneurs create new growth curves, and their improvements and managers exploit the growth curves. Therefore, the prosperity of a nation is dependent on the class of innovative entrepreneurs. But not everyone can be a Steve Jobs; also other entrepreneurs less famous and with less capabilities are needed to foster innovations. Besides innovative entrepreneurs, already setting up an own business without any innovative content instead of becoming a manager drives the economy. It is this business mentality and the mentality of curiosity and of venturing which are needed for value conversion. Therefore, entrepreneurship is crucial for any society. The Chinese possess the drive of being an entrepreneur more than any other country. To say it popularly, every Chinese wants to have his own business and be his own boss now or later. This entrepreneurial drive and mentality are much more a danger for the rivalry between the USA, the EU, and Green China than any R&D program.

In the last decade, the notion, or better buzzword knowledge economy, became popular in governmental circles, the EU commission, and academic world. But the knowledge economy is an empty buzzword. Every industry even the simple ones require knowledge and experience. Not only atomic engineers and investments bankers require knowledge and experience but also bakers, barbers, and welders need education, knowledge, and experience to excel in their job. There are millions of hamburger bakers over the world, but only one came up with the idea to start a hamburger chain. Therefore, the knowledge economy is the wrong notion; it should be the entrepreneurial economy (van Someren 1991, 2005). Manpower combined with entrepreneurial spirit fuels growth and development and not knowledge as such.

When looking at the growth curve and the creation of the new sustainability growth curve, we know two things. Europe is at the end of their industrial growth curve and should make a jump to the new growth curve. The Europeans mainly see the end of the growth curve, and they try to keep and conserve the accumulated wealth and way of doing business, for example, protecting the oil business instead of investing in renewables. Defensive strategies and doing the same in other regions such as China characterize the end of the growth curve.

China started their jump by first earning money from an improvement of the old growth curve, being the outsourcing improvement, but is now focusing on the new sustainable growth curve. In the beginning of such a growth curve, you look up the curve, and you see no end; the sky is the limit. The combination of entrepreneurial mentality and their bias to betting, the perceived unlimited opportunities, drives many Chinese entrepreneurs into gambling. Instead of making strategic plans, they pick up any opportunity and try many things without any connection. But Chinese also know what goes up must go down; therefore, there will come a point in time when they have to make choices and make a strategic plan.

Well planning is not new to Chinese leaders. The next section deals with strategic planning the Chinese way; in particular, the 11th and 12th five-year plans will be summarized from a Green China perspective. More important than the facts and figures is how to deal with these core planning instruments of Chinese government. For this interpretation, the previous sections of this chapter come into play to increase your chances to profit from the Chinese plans. But first, the big policies.

3.4 The Yellow Dragon with the Green Blood

3.4.1 The 11th Five-Year Plan

If we want to understand the future, we have to understand the past. Most Asiatic readers understand this, but for the majority of Western readers, this is disappointing news. Westerners think the future can be created without the past. This is not true, and therefore, we will put the 12th five-year plan in the light of the 11th five-year plan. The reason is that the core issues of China's future are dependent on two central issues: creating a sustainable society and becoming an innovating society. A true sustainable society can only be created by innovations, and innovations are also needed to realize any other future plan of the government and businesses. Both issues of innovation and sustainability have already been addressed in the 11th five-year plan. It comes back again in other policies and the 12th five-year plan. It is obvious that the Chinese government has not been able to solve these issues neither in the previous plan nor in any other governmental policy.

As we will see, like in the USA and the EU, over the years, the innovation issue has been tackled from different angles. But the success stories are only in certain sectors or specific cases. Every society tries to cope with all knowable and expected developments in the short and long terms. Governments, institutions, NGOs, firms, and individuals formulate policies and strategies to reduce the uncertainty of the future and to create a path toward the future. China has the tradition of making five-year plans to give direction to the developments in the short and long terms.

Understanding the future developments and identifying future opportunities in China requires to have knowledge of the plans of the central government. These five-year plans give direction to investments, projects, and ambitions of institutions and firms involved in realizing the goals of the policy makers. For Western companies, it shows which the main markets of the future will be.

But not only central government has a big influence on the development of the economy. Also local governments and the governors of the provinces have their own ambitions besides realizing central governmental plans. Provinces differ in many respects from each other due to differences in, for example, history, traditions, cultural values, size, climate zones, economic development, resources, and local demands. In China, regional markets and its needs differ substantially from each other. Therefore, regional policies differ accordingly. In fact, on many issues, the provinces have a tradition of being each other's rivals. This internal competition, for example, shows up when the central government makes plans for selecting regions as economic growth zones or when it is tried to acquire foreign investments. This combination of central and local plans gives many indications, along which strategic lines they want to develop their economy in the near and longer future.

For being knowable about China's future, two documents are important. The 11th five-year plan as well as a white paper on "China's Peaceful Development" published by the Information Office of the State Council has already given early insight into the strategic direction of China. Both documents reveal the contents, aims, and direction of China's emulation process. For both Chinese and Western companies, it is relevant to know what China's government is expecting from them. But more is to follow. The notion of innovation has been mentioned now in three consecutive five-year plans showing that the issue cannot be tackled within one period but needs long-term support. Western countries are not doing better; innovation is very difficult to grasp. We will see that the West and China follow different paths with different results. But first, we look what were the plans and what is ahead in China. Many of the old ambitions have not been realized yet, and some are reformulated. Here, two big policy traps lie ahead: old wine in new bags and a forward flight.

The central government of China is of course aware of the dynamics of the changes since opening up the market with Deng's quest for prosperity. Opening up markets in similar situations like Russia and former East Germany show that there are no recipes and the outcome of any measure, despite large sums of money invested, is very unsure due to market dynamics. Therefore, a balance between central planning and giving opportunity to individual companies within the conditions of international markets and internal regulations has to be sought. This holds especially for such a large country like China. Generally, these five-year plans show the ambitions of the central government and frankly speaking the points where China needs improvement.

The 11th five-year plan covered the time period of 2006 till 2010. China's premier, Wen Jiabao, characterizes the previous plan as follows:[3] "Domestically, there are many favorable conditions as well as a number of restraining factors and difficulties." To observers, what Wen meant was that opportunities and challenges for the central leadership exist. Xia Xueluan, a professor of sociology at Beijing University, says, "The coming five years will be a golden opportunity for the central leadership to lead the Chinese people toward prosperity and will test its governance capacity as well."

Compared to previous five-year plans like the household contract responsibility system introduced to the Chinese countryside in the 1980s and the socialist market economy of the 1990s, the previous plan heralds a major shift in the ruling party's policy toward green GDP and sustainable development. The fast-growing economy, heavy energy consumption, and wide gap between the rich and the poor have been

---

[3] Citations from www.gov.cn, "New 5-year plan embodies opportunities and challenges for China's leadership" 06.03.06.

included in Wen's report as issues to be tackled in the coming five years, alongside goals to improve the country's economic structure, innovation capacity, and growth pattern; curb pollution; create more jobs; balance investment and consumption; reduce income gap; and bolster public welfare. "We should seize opportunities, exploit favorable conditions while avoiding dangers, remain vigilant against adversity, properly handle risks and challenges, and press ahead with reform and opening up and the modernization drive," Wen said in the report. Now what does it mean?

At the time of the revolution, the economy was still dominated by agriculture; in 1952, about 60 % of GDP was generated by the agricultural (primary) sector. Both the first and the second periods were dominated by industrialization, rather than "post-industrialization" that took place after World War II in developed and most less developed countries. As a result, China ends up being highly "industrialized" by the end of the century. In 2003, the GDP structure of China is 12.5, 46, and 41.5 in terms of percentage for the primary, secondary, and tertiary industry, respectively.[4] The growth in manufacturing and the relative shrinkage of agriculture went on also in the 1990s, and the value-added share of the service sectors remained almost unchanged until the second half of the 1990s.

The 11th five-year plan addresses a range of problems.[5] In 2005, the country's GDP reached 18.23 trillion RMB, an increase of 9.9 % over the previous year. According to the plan, the government will be in a better position to address citizens' problems when the GDP reaches $3.2 trillion in 2010. By that time, the nation's per capita GDP will be around $2.400, compared with $1.700 in 2005. Ma Kai, minister of the National Development and Reform Commission, said that China's average urban per capita disposable income in 2010 would be 13,390 RMB ($1.650), while rural per capita net income will be 4,150 RMB ($511). But the gap between urban and rural income levels will still be as wide as it is now, despite the various policies being proposed to help the development of the countryside.

Also, the central government plans to extend more financial support to improve rural drinking water supply, aiming to provide safe drinking water to 100 million residents by 2010. Premier Wen Jiabao proposed an annual growth rate of 7.5 % for the national economy from 2006 to 2010. In the predicted GDP growth between 8.8 % and 9 % in the past 5 years, the losses as a result of huge wasting and pollution have taken up more than 2 % of the GDP. According to the estimation of the United Nations Environment Program (UNEP), China's economic losses due to its deteriorating environment account for 8 % of the nation's GDP each year.[6]

---

[4] ShulinGu and Bengt-ÅkeLundvall, 2006, China's Innovation System and the Move Toward Harmonious Growth and Endogenous Innovation,in: ShulinGu&Mark Dodgson (ed.), A special issue of Innovation: Management, Policy& Practice, April 2006.

[5] Citations from www.gov.cn, "Five-year plan addresses pressing problems," 07.03.06 and People's Daily Online08.03.06 and www.gov.cn "Resources conservation policy listed in five-year plan," 13.03.06.

[6] www.gov.cn, "China adopts various regional growth modes," 10.03.06.

The livelihood of both urban and rural residents will see improvements in terms of housing, transportation, education, culture, health, and environment. According to Bai Hejin, one of the experts responsible for drafting the guidelines and former head of the Macro-economy Research Institute under the National Development and Reform Commission (NDRC), "There is no need to damage environment for a high growth…in the coming years, if we can cut environmental pollution by one percentage point, then the value created out of the 7.5 percent growth rate will stand equivalent to that out of 8.5 percent. Moreover, people will enjoy more benefit with improved environment."

China expects remarkable progress in energy consumption efficiency. Energy consumption per unit GDP should be about 20 % less than before and water consumption per unit industrial added value down by 30 %. The coefficient for water use efficiency in irrigation is to be raised to 0.5 and the comprehensive utilization rate of solid wastes to 60 %. "China is a big country of resources in terms of total reserves, but it is a small one if its population of 1.3 billion is taken into consideration," said Wang Shusen, a deputy to the Tenth National People's Congress (NPC), China's top legislature. Official figures show that China's per capita possession of freshwater is only one fourth of the world average; arable land, less than 40 %; proven coal reserves, 62 %; proven oil reserves, 7 %; natural gas, 4.5 %; and forest, one fifth of the world average. Furthermore, the per capita possession of 45 kinds of mineral resources is less than half of the world average. A survey released by the Chinese Academy of Sciences shows that China ranks among the world's most wasteful users of natural resources. The survey, which listed China 54 out of 59 countries, proves that China does not fundamentally break away from its economic growth model that relies on the "intensive use" of natural resources and energy, says the report.

The economic development in every region should focus on the prerequisite of saving energy. This is also in line with the country's development of a circular economy.

With regard to innovation, the following issues are addressed.[7] China, which boasts the four ancient inventions of paper, gunpowder, compass, and typography, is now striving to re-embrace its past glory in a matter of 15 years, aiming to turn itself from a manufacture center to a major inventor of the world.

To improve the nation's innovation capacity, it was listed on the top agenda of China's 11th five-year plan. Premier Wen Jiabao pledged, while delivering a report on the work of the government at the opening meeting of the NPC session, that China will speed up building an innovation-oriented country in the coming years and launch a batch of technological projects covering the information technology, biotechnology, environment, and energy. In 2005, the country's expenditure on research and development only accounted for 1.3 % of its GDP, 0.2 percentage points lower than the expected 1.5 %. The gap demonstrates a lack of enthusiasm of

[7] Citations from www.gov.cn, "China seeks impetus from innovation," 14.03.06.

enterprises for innovation and the insufficiency of technological investment of the whole country, said Guo Guoqing, a professor of the People's University of China and member of the National Committee of the Chinese People's Political Consultative Conference, the country's top advisory body. By the way, 1.3 % is in the same range as the average percentage of GDP spent on R&D in the EU member states but less than the almost 3 % in the USA. In the Mid-to-Long-Term Scientific and Technological Development (2006–2020) (MLP), the aim is to spend 2.5 % of GDP on R&D in 2020. Therefore, when China is able to reach its targets and to increase the percentage spent on R&D, the EU will get a big problem and lag behind when no improvement can be demonstrated.

But it is not only R&D outlays and number of patents which pave the way to success. More relevant are the nontechnical aspects of innovation and the ability to create products, services, and new value creation opportunities. This challenge is far greater than allocating money. A deputy of NPC Mr. Liu Yang said, "money itself is not enough to bolster innovation…The nation needs to cultivate an innovation-minded concept and only with such an environment can innovation become a major driving force to the country's development." Also with regard to sustainable development, Mr. Yang Yang Zhenchao, an NPC deputy and director of Anhui Economic Committee in east China, remarks, "efforts have to be made to develop high-added value, hi-tech and high-yielding industries, promote the use of energy-saving equipment and technologies, make good use of government policies, and enhance the law enforcement capabilities of environmental protection departments."

The 11th five-year plan differs greatly from previous five-year plans in two respects. First, central policy making will not be homogenous for every region in China but where relevant differ accordingly to the needs and differences per region. Under the proposed program, China calls for different economic growth modes for regions based on environmental characteristics rather than on a unitary mode.[8] "This is the first time for China to propose different modes of development for different regions in a five-year development program…This is a new concept that is extremely significant…This will help fairly allocate public resources and solve specific problems that are unique to a particular region," said Chen Jianhua, secretary of the Gannan Tibetan Autonomous Prefecture Committee of the Communist Party of China (CPC). Li Shantong, a member of the National Committee of Chinese People's Political Consultative Conference (CCPC), said the former regional division policy is one reason for the broadening gap between the economically developed regions and China's Western environmentally sensitive areas. "It was hard to take advantage of what the different regions have to offer due to a lack of cooperation among them," said Li.

Secondly, with regard to the economic development, not only purely economic issues get a central place but the balance between economic, social,

[8] From www.gov.cn "China adopts various regional growth modes," 10.03.06.

Table 3.7 CO_2 emissions and ecological footprint

	CO_2 emission in mio tons (2004)	Per inhabitant in ton	Footprint in mio hectare (2002)	Per inhabitant	As percentage of total country size (%)
China	1.021	0.8	2.049	1.6	201
India	301	0.3	784	0.8	210
USA	1.616	5.5	2.810	9.7	205
Europe	955	2.5	2.164	4.7	207
Japan	338	2.7	544	4.8	569

Source: NRC Handelsblad, 14.01.06

and ecological issues are seen as belonging to the single issue of progress and prosperity. However, this way of thinking—balancing seemingly contra dictionary elements—is a key value of Chinese philosophy and is therefore not a surprise. The ratio for this new foundation of the 11th five-year plan is the accumulated effects of the opening up markets for China as a whole and the different local dynamics of the (regional) economy. The accumulated effects of China in the past decades show on the one hand unrivaled economic growth rates based on manufacturing but on the other hand a widening of income levels and prosperity between rich coastal regions and the traditional rural areas as well as detrimental ecological effects. These developments on the long term do not contribute to the stability of China. Taking care of stability and harmonious progress has always been the core task of Chinese leaders, and it represents their reason of existence.

The 11th five-year plan shows that Chinese leaders are aware that they are in a crucial phase, and measures have to be taken in order to continue their development. These measures boil down to two main categories: endogenous innovation and harmonious development. Endogenous innovation implies that China should be able to create innovations by themselves instead of being and staying dependent on foreign knowledge. Harmonious development means the introduction of sustainable development at the core of future progress. Taken together, these new starting points for the coming years shift the attention from "getting rich first" to "common prosperity" and from "growth rate" to "sustainable development." The growing necessity for sustainable development at the time of the 11th five-year plan shows Table 3.7 and Fig. 3.2.

The CO_2 emissions in China belong together with the USA to the highest in the world. The ecological footprint—measuring the number of acres of land needed to fulfill the consumption pattern—is also belonging to the highest. When sustainability is measured along ecological footprint and social inequality leading to potential economic instability, Fig. 3.2 demonstrates that action is required.

The social inequality is measured with help of the Gini coefficient. With regard to China, already around 2004, China has passed the threshold level of the danger zone where inequality can lead to instability. The latter is the biggest fear for Chinese central government. For this reason, sustainability became again a central focal point in the successor of the 11th five-year plan.

Fig. 3.2 Sustainable development. (Source: World Bank data bank Gini-coefficient, CCICED-WWF, Report on Ecological Footprint China)

3.4.2 The 12th Five-Year Plan

The 12th five-year plan is a continuation of the previous plan but simultaneously takes a different route. In Table 3.8, a summarized overview of the two plans is presented.

The official GDP growth rate is now 7 %, only 0.5 % lower than the previous target. However, the difference is that the focus is now shifting from pure growth mode to a long-term prosperity growth mode. The verdict of the pure growth mode has been announced around 2007 when Premier Wen Jiabao characterized the growth model as the four "Uns": unstable, unbalanced, uncoordinated, and ultimately unsustainable. The break-neck, winner-takes-all, and no-matter-the-cost pursuit of growth has served out, and social and ecological dimensions will have to be balanced against economic dimension.

This implies that the notion of harmonious growth pointing to Triple P growth has not worked out in practice. The notion of harmonious growth has been replaced by inclusive growth meaning that GDP growth should be replaced by Triple P growth taking into account economic, social, and ecological aspects.

Translated into policies, the 12th five-year plan has the objective to close the urban–rural divide and close the gap between rich and poor. The latter is connected to the shift of development attention from the eastern coast regions toward western regions of the country. Also in the near future, the inland consumptions get attention besides export which has been the economic engine of China's success

Table 3.8 Triple P key themes: 11th and 12th five-year plans

Triple P dimension	11th five-year plan	12th five-year plan
Economical	GDP growth rate 7.5 %	GDP growth rate 7 %
	From unitary to regional differences growth mode	From growth mode to long-term prosperity
	Harmonious development	Globalization
	Endogenous innovation	Inclusive growth
	Circular economy	Indigenous innovation
	Foster technical innovation	From export to domestic consumption
	Support 11 sectors	Develop service industry
		Nine key industries
		Support of seven emerging industries
Ecological	Reduction of energy use per unit of GDP	More mandatory green targets and green development indicators for keeping officials accountable
	Reduction water consumption	Energy saving, promote renewables
	Reduction of major discharge pollutants	
Social	Educational programs	Close urban–rural divide
	Start with social security and pension plans	Close income gap and reduce inequality
	Fostering middle-class development	From eastern to western regional development
		Well-being/happiness of Chinese people

till now. In order to avoid own interpretations of targets and the relevance of greening China, the targets related to sustainability are mandatory, and officials will be held accountable for results.

Another interesting aspect is that before opening of Congress 2011, Wen Jiabao mentioned that the task of the Party is to make Chinese people happier. The well-being of the Chinese people is a core issue for the party in order not to lose its legitimacy in the eyes of the Chinese people. Illustration of these dangers is inflation, high food prices, high housing costs, social security, and too high income discrepancies between regions and population segments. According to a survey conducted by people.com.cn., the poll found that among more than 95,300 respondents, 95 % believed the current pension system is unreasonable and many suggested a quick reform to ensure its equity and fairness. China's dual pension system, introduced in the early 1990s, allows civil servants to enjoy higher pensions than enterprise employees do after retirement.[9] Sustainable notions of harmonious development, circular economy, endogenous innovation, and GDP growth used in the previous plan period are now being replaced by notions like inclusive growth, indigenous innovation, and industries and regions.

[9] China Daily 03/02/2011 p. 3.

The 12th five-year plan pays special attention to fostering the service industry as the new growth sector and future powerhouse. This focus is not coincidental or at random. An important reason is that the growth of the service sector in the past years lags behind compared to other sectors and other developed countries. Another reason is that the service industry has become the key source of foreign direct investment (FDI) in China.

China has lost the rat race in several technology areas like fossil fuel vehicle sector, and it is now trying to catch up early in service industry with the aim not to lose this battlefield to the West. According to the Ministry of Commerce, the service sector accounted for 46.8 % of the nation's FDI as a whole, up 3 percentage points from the previous year. "In addition to luring foreign investment, the service industry's fast growth will depend on the development of domestic small- and medium-sized enterprises, something that needs government support in terms of tax and fund raising channels," said Xia Jiechang. Another reason is that service industry offers a lot of opportunities for SMEs and employment.

The development of the manufacturing sectors is now coming into the phase of capital-intensive and labor-saving equipment. As a consequence, in the coming years, the manufacturing industry will not be capable to employ the majority of Chinese labor force. The Chinese production mode is not sustainable from the perspective of rising labor costs, ecological detrimental effects, and social inequality. The labor costs in China rise with 15 % till 20 %, and the difference with the USA will reduce from 55 % to 39 %. In case of Europe, in 2010, Chinese labor costs are 25 % of European level, and in 2015, it is expected that it will rise to 38 %.[10]

But also other costs such as transport, quality, automation, ecological measures, and energy rise. For example, the Taiwanese electronic supplier Foxconn gives the traditional answer of increasing automation in factories to reduce labor costs.[11] The service industry offers a way out. The development of large-scale transactions-intensive industries such as the wholesale and retail trade, domestic transport and supply chain logistics, health care, and leisure and hospitality have the potential of becoming a new growth and employment engine.

Moreover, four other developments and goals fit into developing the service industry: increasing urbanization; decreasing the dependency and intensity on natural resources; increasing wages for enabling increased domestic consumption; and the development of a social plans like education, security system, pensions, medical care, culture, and sports. The aim is to increase the contribution of the service sector to the gross domestic product (GDP) by 4.2 percentage points by 2015 to around 48 %. The number of people employed in the sector is expected to grow by 5 percentage points to 41 %. "Value-added tax is levied only on manufacturing, while services are subject to a business tax that raises a much smaller amount. Given the heavy dependence of local governments on value-added tax revenue, they have a strong incentive to promote manufacturing rather

[10] BCG, August 2011, Made in America Again, Why manufacturing will return to the US.

[11] FD, 07.11.11, Foxconn omzeilt hoger loon met robots.

than services," said the Asian Development Bank in a report. "The employment-intensive domestic sector can be the turning point in the government's efforts to solve the dilemma of restructuring the economy and ensuring rapid growth to maintain employment levels," said Li Wei, economist at Standard Chartered Bank.[12] This almost inevitable cause of direction will also hit old institutional practices like the household registration system called the hukou. The hukou system hinders the emergency of a flexible labor market which is absolutely necessary to ensure a healthy service sector.

Paying attention to the service industry in a highly technology-adoring and biased country can be called a watershed in China policy making. In China's culture, hard facts and concrete physical products get more attention and are regarded to be more important than soft facts and service-based offerings. Is technology then abandoned? No absolutely not; on the contrary, technical innovations and large infrastructure projects still dominate the 12th five-year plan. On the one hand, this is caused by the nature of the new industries dominating a sustainable economy. On the other hand, the relevance of nontechnical innovation for transforming from an unsustainable toward a sustainable society is still not completely recognized and acknowledged. The latter subject will be discussed later; we now focus on the other core issues of the 12th five-year plan. In this respect, seven priority industries have been named which will form the new basis of the future Chinese society. These seven industries have all a direct or indirect link with sustainability. In Table 3.9, these industries are mentioned, and a few potential opportunities in the field of sustainable development are added.

As can be seen from Table 3.9, the targeted seven strategic emerging industries are supposed to be dependent on new technologies and new infrastructure. In this respect, the 12th five-year plan is not different from other five-year plans or most other governmental policies. Policies are dominated by new technology, large new infrastructure plans, and breathtaking individual projects. The seven strategic emerging industries together with the service industry in the five-year plans can be seen as a bold initiative to change the characteristics of the Chinese economy.

With regard to the new direction of China's economy, the aim of expanding the business of state-owned and private firms abroad is another bold feature. Globalization will not only be important for securing the necessary natural resources needed by the manufacturing industry and fueling the economy but also for global presence of Chinese firms in intermediary and end consumer markets. Establishing Chinese firms in the global top 1,000 and Fortune 500 list is an unofficial but clear aim of the government. According to Shao Ning, vice-chairman of the State-owned Assets Supervision and Administration Commission (SASAC), "Central SOEs should concentrate on completing their industry chains, increasing market share and developing into top-ranking enterprises in the international market during the 12th five-year plan (2011–2015)."[13] This remark should be seen in the context of

[12] China Daily 03/02/2011 p. 7.

[13] China Daily 02/25/2011 p. 13.

Table 3.9 12th five-year plan: Seven strategic emerging industries and its sustainable opportunities (examples)

12th five-year plan: seven strategic emerging industries and its sustainable opportunities (examples)	
1	**New energy**
	Wind, solar, nuclear power, clean coal, natural gas pipelines, shale gas, oil equipment, nuclear safety, offshore wind, and biomass
2	**Energy conservation and environmental protection**
	Recycling, pollution emission equipment, window energy equipment, water supply, waste water equipment, and irrigation
3	**New energy vehicles**
	EV, PHEV, HEV, BEV, energy efficient vehicles, global markets, and batteries
4	**Biotechnology**
	Bioenergy, drugs, and new medicines
5	**New materials**
	Rare earth materials, building materials, LED lights, modular building, zero-energy buildings, and eco-cities
6	**High-end equipment manufacturing**
	High-speed railway equipment, aerospace, and telecommunications
7	**New IT**
	Network infrastructure, smart grid, smart metering, and Internet security

the aim mentioned in The guiding principles of the program for Mid-to-Long-Term Scientific and Technological Development (2006–2020) (MLP) to become more independent of Western technologies and getting a larger stake in high-end part of the value chain.

Besides industries and globalization, the development of a national innovations system remains a priority. Here, the aim is to connect research on universities and knowledge institutes with the business sector. Also here, in the previous decades, the initiative for innovation has been at governmental institutions. With the new five-year plan, both state-owned and private firms are supposed to take the lead allowing the market to do its work. Of course, government will set priorities and give direction especially in selected industries.

Now it can be seen what greening China means: not only green development, green cooperation, and green science and technology but also creating world-class leading and innovative firms and industries. However, in this book, the focus is on the greening of industry.

3.4.3 The Long March from Interpretation of Five-Year Plans to Business

The five-year plans look like the golden mountains mentioned in Chap. 2. The enormous investment opportunities and growth potential on paper look like one big opportunity. But for Western firms, golden mountains do not exist. If Westerners think they can take away a part of this cake, they are wrong. Of course, Western firms have competences, knowledge, and experience leading to opportunities in all seven Chinese strategic emerging industries.

But in order to explore and exploit these opportunities, the pure Western interpretation of grasping market opportunities should be abandoned. One of the threats to traditional Western market penetration and market development is that the Chinese will have copied green technology before the Westerners can earn money of it. Another threat is that only the exploitation and selling of the technology and staying ahead of the Chinese competition are put at the core of the business development of Western firms.

The next chapter will show typical cases of Western misinterpretation of opportunities and threats. These misinterpretations and behavior work contra-productive to the technology and nontechnical competences the West has to offer. The issue at stake cannot only be reduced to cultural differences. It is about the complete setting of the society and how central and local government work, the role of public-owned and private organizations, the market structure, and market behavior. Our core point here is that strategic innovations are needed to align Western technological advance and nontechnical innovations with Chinese market circumstances, market structure, and market behavior.

References

Frank, A. G. (1998). *ReOrient: Global economy in the Asian age*. University of California Press. Berkeley, Los Angeles, London.

Elvin, M. (1973). The Pattern of the Chinese Past. Stanford Unievrsity Press. London.

Jong, H. W. de. (1985). Dynamische Markttheorie, Stenfert Kroese.

Someren, T. C. R. van. (1991). Innovatie, Emulatie en Tijd. De rol van de organisatorische vernieuwingen in het economische proces, Tinbergen Institute Research Series, no. 9, June 1991 (diss.).

Someren, T. C. R. van. (2005). *Strategische Innovationen. So machen Sie Ihr Unternehmen einzigartig*. Wiesbaden: Gabler Verlag.

Doing Business in China: Success and Failure on a Cutting Edge

4

Carried or Capsized by the Wave

Many are interested in the promising golden mountain in the greening China. But only a few can really make money. In the years when we are trying to help the European companies and governments to optimize their China operations, we see unfortunately too many failures. In this chapter, we will tell some of the true stories which can help us to learn from them. Because of the concurrent sensitivity and the privacy, the names of the organizations and persons here are fictive. These cases are illustrative for many organizations trying to do business in China. To our experience, even multinationals active over decades in China sometimes show behavior you would not expect from such diehards. The reason is that these experienced organizations know on paper what to do. But when the paper strategies are carried out, most organizations do not know how to deal and to develop the measures further. In this chapter, a few cases from our practice demonstrate the manifold and different issues from operational to strategic level in dealing with Chinese. They all show the all-or-nothing results of China business. Deliberately, we have chosen not only success stories but cases which are common in China and all can learn from.

4.1 Many Ways to Capsize or to Be Carried by the Wave

Western managers are increasingly aware of the different ways of doing business in China. Some entrepreneurs are extremely successful even without deep knowledge of Chinese society. They just do it their way. But they are the exception. Most SMEs and MNEs have their difficulties, and they do not want to admit it. It is the Western way of losing face, for example, appointing Chinese as leader of your China practice instead of an expat is common in multinational companies. But this does not give you the guarantees that from now on, the golden mountains will appear. From the view of the headquarters, still lack of knowledge exists about the

T.C.R. van Someren and S. van Someren-Wang, *Green China*,
Management for Professionals, DOI 10.1007/978-3-642-28810-4_4,

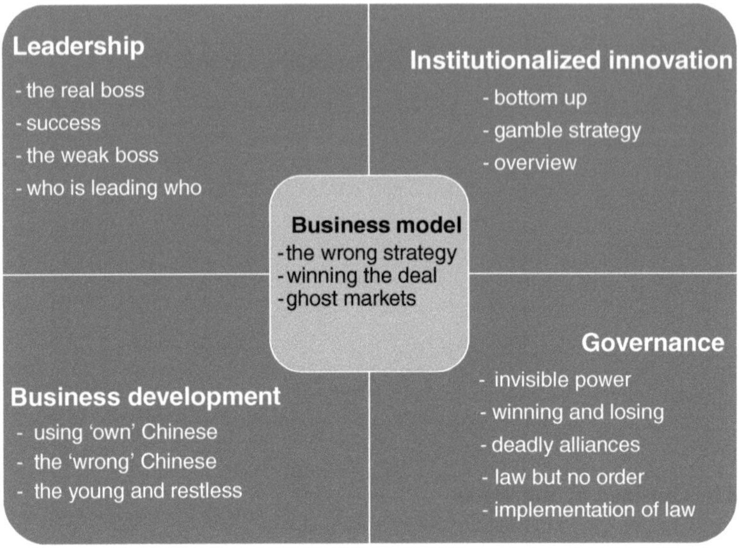

Fig. 4.1 Success and failure on all levels

right selection, how to lead your Chinese counterpart, how to change strategies according to headquarter guidelines, and how to deal with the growing power of the Chinese branch.

In order to organize this chapter, we categorize the cases according to five dimensions: leadership, business development, business model innovation, institutionalized innovation, and governance. Figure 4.1 gives an overview, and this format comes back in later chapters.

The number of cases is endless, and therefore, we only give one or two examples for each category mentioned in Fig. 4.1.

4.2 Leadership

Libraries full of literature exist about leadership, and we are still looking for the holy grail of leadership. And now, we introduce an additional aspect of leadership: how to deal with Chinese leadership. Is there a Chinese way of leadership? And if yes, in what respect does it differ from Western views on leadership? In Table 4.1, a few illustrations show that Chinese leadership indeed exists and has some profound differences with Western leadership styles.

According to a Chinese legend, if you want to catch the dragon, you have to catch the head of the dragon. Since the rulers of China found out that Confucianism emphasize hierarchy and order, the Chinese society was ruled in the same hieratical order for thousands of years. Leaders and their leadership quality have caused the raise and fail of different dynasties. The performance of a Chinese organization also

Table 4.1 Leadership issues in The West and China

Leadership issues	Western	Chinese
Strategy	Create shareholder or stakeholder value	Create network (guanxi) value
	Short-term clear aim and direction	Long-term big picture vision
	Business model	Business tactic
	Inconsistence avoidance	Paradox synthesis
	Position	Battles
Business development	Internal competence and resource-based growth	Internal resources and harmony-based growth
	External M&A-based growth	External network and opportunity-based growth
Value chain management	Optimal clients and markets	Maximal markets and number of clients
	Sourcing to enforce core business	Sourcing to extend core business
Partner management	Alliance management	Network management
Innovation	R&D	Search and get and development
	Knowledge management	Resource management
	Top down and bottom up	Top down and bottom up by order
HRM	Recruit the best qualified	Recruit the loyalist or with the best guanxi
	Empower with organization chart and planning	Empower with hierarchy, trust, and reward
	Retention using contract with clear reward and punishment	Retention with "family" or group forming and shaming
Performance	Short-term bottom line	Long-term benefit
	Financial target	Financial power
Operation	Manage with procedures	Manage with examples
Management style	Organizer with expertise	Charismatic strong father

mostly depends on the quality of the leader. Many Western companies failed to earn money in China when the external conditions are good. In many cases, you can trace the failure back to failing managers in China and/or unsuitable leadership in the headquarters.

4.2.1 The Headless Chicken

This is a story of the chicken with too many heads but in fact "no head" in the Chinese eyes. An environment instrument producer wanted to sell their instrument in China. It is a hundred-year-old family company with a solid history. They have focused on innovation for years and possess the state-of-the-art instruments. They have worldwide network of sales agents to help them became one of the biggest in their field. They have set up a rep office in China for several years. Despite the big investment of sending expensive expats and building an agent network, they still cannot see the figures from Chinese office turn black. During a visit of a delegation, a meeting was organized with the sales manager in Chinese office. He came to us with the question if we could help them find the way to improve the sales in China. Of course, first, he had to discuss it with his boss in his home country about it. When we called him a month later, he still did not know if they wanted to do that. The family which owns the most shares wanted to stop the money-losing Chinese office. The CEO was new. He did not want to offend the shareholder but also did not want to lose the golden opportunity in China. The headquarter sales manager wanted to have a worldwide uniformed sales strategy. The Dutch manager in China favored another Dutch consultant who has an office near to their office in Beijing. The Chinese manager in Beijing wanted to hire us but did not want to take the responsibility. In short, everyone wants to make the decision. And no one could really show the strong leadership which is necessary for the Chinese market. No wonder they are not doing well in China.

4.2.2 The Strong Head that Cannot Turn

Very often, we work with capable top managers in China. Different than most European managers, they mostly have very strong ideas about their company strategy. Strong does not necessarily mean better or formulated along the newest insights of strategic handbooks or supported by most famous strategy consultant practices. Very often, Chinese strategies are in the eyes of many Western managers primitive and not state of the art at all. But the perceived drawback of a failing state-of-the-art strategy formulation is often compensated by a great flexibility and uncompromised perseverance in achieving the long-term goal. This combination is often deadly for Western rivals and beats handbook like corporate strategies. But in some cases, the leadership is so concentrated in the top level or even on single person that operational managers cannot give input to new strategy even when it is necessary. One of our cases is from a Chinese solar cell company. In 2005, we meet their sales manager in Europe when they are looking for selling channels. They had products with good quality and price. But technology of the products was of a previous generation. We warned them about the risk of relying on the competitiveness of lower price. If they went on the same way, they will be replaced by others with better technology linked with better quality. The only way to sustain their business is to develop a good working strategic agenda for Europe. Facing the upcoming competition from other solar cell companies, the sales manager agreed with us. He wanted to use our influence to show his boss that they need to do business differently in Europe. We met the big boss. When we talked about developing their strategic agenda with the input of his operational managers, he laughed. He waved away the idea with his hand. We got an hour lecture from him about how he worked himself up in the army, managed a state-owned company, and become multimillionaire by exporting low-tech and low-price products to Europe. He was successful indeed and had all the rights to be proud on his achievements. He was sure that his investment in new energy sector will be successful too. He gave us his big picture of the company controlling from the whole product chain of the solar cell. He was right for several years. From 2005 until 2011, they have only seen growth with double numbers. But in 2011, they are balancing at the edge of bankrupting. Their products are wiped away by the new-generation solar cells with much better results.

4.2.3 The Boss Cannot Be Weak

To be a good boss in China, you will have to act like the sun. That means to give warmth to the people who rely on you and to be the example to the people working for you. Showing weakness is not a wise thing to do. Many Western leaders cannot understand this. For many years, we have served many Dutch leaders. Most of them are very willing to cooperate with Chinese. Their ways of showing it have sometimes the opposite effect. For example, to prepare the trip to China, they mostly got informed about Chinese manners. One of the things they read is that they should use both their hands to hand over their name cards. This is true in the case of name card exchange between equals. But it is not wise to do it to everyone. If you do that to the people who are much lower positioned, they will be embarrassed, and your position will be undermined. Another example is the drinking ritual. The advice in most books is to toast and to lower your glass to make it lower than the counter-partner, which is also not true in most cases when the Chinese are not higher positioned like you. Another example is to try to speak English when English is not your strong language. Sometimes, you have very embarrassing situation when business leaders try to read a speech in very bad English which the Chinese interpreter could not understand. It is even worse when the Chinese make speeches without paper and in Chinese. The opposite also happens. One of the leaders we know is very good because he knows how to turn his weakness into strength. Because his eyes are not very good, he decided to use his mother language to speak in his trip to China. He is an excellent speaker without written speeches. That makes it possible for him to look at the Chinese when he spoke. The Chinese were impressed by his charisma because he used his mother language with dignity. Because there are very few people who can speak Chinese and any European native language at a high level, on this occasion, we were asked to be the interpreter. In Confucian tradition, our doctor titles also helped. In China, if a leader is surrounded by capable people, he is also seen as an even more capable leader.

4.2.4 The Turtle with Dinosaur Head

A few years ago, the top management of a famous energy company was not satisfied about their bottom-line results of their China office. In their very long history, most of the time, they are state owned. The organization is used to a culture with strong management control. Most of the managers think that they need the detailed control to get things done. A lot of time is wasted in reporting and controlling from different levels. When they opened their business unit in China, they could not keep any of the Chinese managers. Most of the Chinese managers think that they are controlled because the headquarters does not trust them. They feel face losing in front of their people every time they are controlled. The company decided to hire a retired Dutch manager who has been working for a very international-orientated Fortune 500 company for years in China. The thought was very good. An old man with a lot of experience should be an asset in China. Many Chinese organizations hire retired people to work for them for the expertise and/or the network. The older-generation Chinese do not have hobbies except work. They are most of the time very grateful to an organization which make them feel needed. They would do their best to help this organization. But a retired Westerner is different from a retired Chinese. They usually do not want to work longer than necessary. When they do have to work longer, there are always strong financial motives. In this case, the old Dutch manager had a very expensive divorce and an expensive young Chinese wife. He has to work. His motive to work for this company is not the long-term benefit but his own two-year plan. After 2 years working as an expat for this well-paying company, his shortage in pension payments will be compensated enough by the extra 2-year income. He made very good impression by the headquarters because he is good at reporting and corporate speaking. But the headquarters did not know that he has a very negative network. In the years when he worked in China, he learned the wrong things and made a lot of enemies. Many people who worked with him did not like him. On top of that, he has a very wrong idea about Chinese networking. He considers it a way of corruption. So he used his position to enrich himself when he worked for the multinational. But he forgot to share the benefit with others. The latter is very deadly in Chinese network circles. The Chinese way of networking is far more complicated than gift giving or corruption. In principle, one has to have long-term investment in relationship to create guanxi. It is just like investment in stock market; you have to buy low and sell high, not the other way around. The best investment is in time of difficulty. If you do a Chinese even a small favor, for example, give him face, at the lowest point of his life, he will always remember it and want to return it when he has the capability. Another rule of stock market investment is to spread the risk. One should never only be a friend of one kind of people. Many expats have a very limited social circle. Most of them only get along with Westerners. Some also make friends with Chinese but only those on management level. Only thinking of oneself is the worst way. In the stock market, one is wise to invest for the long term. The Chinese network building also requires long-term perspectives. If one was friendly to a Chinese in a low civil service position 20 years ago, one may have a very powerful friend now.

4.3 Business Development with People

4.3.1 The "Own" Chinese

"Good people surrounding and supporting me is the most important" is the most used phrase to describe their secret of success for the Chinese top managers.

The good one may not be the most qualified one. A European alternative pharmacy company approached us after our presentation for a chamber of commerce about the importance of having good people in China. They became big by importing Chinese traditional medicines first from China. After years of buying from Chinese pharmacy companies and suffering from unstable quality and quantity, they wanted to have their own factory. The factory was being built, but they still do not have good people to work for them. Their "own" Chinese helped them to buy the medicines and later coincidently also found suitable locations near to their own hometowns in China. But the "own" Chinese could not help them find suitable managers throughout the whole year when the factory was built. The pharmaceutical company asked for support. We had contacted our headhunter in China; three candidates were found in a week's time. Those are young Chinese managers who can speak good English and have several years of experience in managing pharmacy factory process, exactly the kind the company was looking for, and they were amazed. They told their Chinese the good news, and they wanted to interview the candidates. One day later, the Chinese partners came with their own list of candidates. They are all coincidently from the same hometown. The managers of the pharmaceutical company asked us what to do. Our answer is that it is wiser to use the Chinese from the same hometown where the factory is built. They may have connections we do not know which may help their business when it is necessary. On the other hand, if they make those people unhappy by not having their people working for them, it may be a disaster.

4.3.2 The "Wrong" Chinese

Due to the hope in lifting with the Chinese economic wave and lack of Chinese employees with a qualification that suits to the Western recruitment criteria, there is shortage of suitable Chinese to hire. In metropolitans like Shanghai and Beijing, the salary level has almost reached the European standard. In remote areas, you have to pay more to attract the "suitable" Chinese. The problem is that if one uses salary as the instrument to get or to keep the Chinese employees, one end up with having the "wrong" ones.

A company in water treatment technology had difficulties to develop their China activities after investing substantially. After investing four million euro in China, they still could not get their foot in the market. As one of the best in treating chemical wastewater, there is nothing wrong with their technology. In their market research based on the official figures of the Chinese chemical industry, a yearly billion-dollar market was estimated. The core question of responsible top management was to find out what was wrong with their China practice. The on-site visit to their Beijing office brought us a in a shiny modern tower in east Beijing. There were four people working in the double-room office. The general manager was a young energetic Chinese graduated from Delft University. The sales manager was a nice-looking young lady from Xinjiang Province. The other two were also very nice young Chinese girls working half as office support half as sales representatives.

They all spoke very good English. Most Western companies like our client prefer it and are willing to pay more to good English-speaking Chinese to get them. During our visit, we found out what was really wrong in the office. None of them have studied in a Chinese university or worked for Chinese organization; no wonder none of them have network in Chinese chemical industry. The girls were not very friendly to each other because they have similar qualification and all wanted to be the sales manager. The most nice-looking one was chosen. That means a lot of energy is used in internal battle. The sales manager from a certain Chinese province has a family member in jail due to political reasons. She was neither happy with the Chinese political system nor glad to deal with the people working in the government. This makes it difficult for her to get a Chinese client. The general manager had put a lot of effort into dining with possible clients in the first year. But the headquarters decided to cut his budget after examining his, at that time, fruitless dining behavior. Since then, he also lost a few of the good leads. None of the Chinese was happy working there anymore. They were all looking for other opportunities. The only reason that they were still there was that no other companies wanted to pay them more.

A company wanted to market their DeNOx technology for coal-fired power plants. They heard that we have useful networks. A proposal was made that we will share their profit if we succeed in getting clients for them. There is a new regulation about the emission last year. There will be a billion-euro market. We spread the news to a few of our connection in the Chinese environment circle. All of our friends who are working for the government, universities, or environment consultancy or engineering companies are excited about this opportunity. They all want to help this company we introduced. Without being paid, because of our connections, some of them helped us talk to the right people from the right organizations. A few promising leads were generated. A big regional coal-fired power plant was introduced to us by the Director General of Provincial Environmental Protection Department. When we first went to the power plant with the director general, although we are the seller, we were received warmly. A banquette was hosted by the number one of the power plant to enforce the newly built connection. In his dinner speech, he thanked us for bringing such famous company with state-of-the-art technology to help them solving the NOx problem. They wanted to be the first to implement it in their region to be an example for other companies. After several visits, technical information needs to be exchanged. We asked the company to send their own people to give the presentations about their technology. After getting the name cards of all the important people of the plant, the European company decided that they did not need us anymore. But when they sent their people without informing us, they could not speak to any of the important people in the power plant. Every time when the DeNOx people tried to call, most of the Chinese leaders were busy having important meetings or not in the plant at all. After several approaches of their own, they came back to ask our help again. But none of our friends would want to help them anymore. In China, it is very unwise to violate one's reputation. If one is proven not to be trustworthy, one will not survive in the market for the long run.

4.3.3 The Young and the Restless

Chinese are very good at turning their weakness into power. One of the famous stories about the great Chinese strategist Zhu Geliang is the Empty City tactic. When he was surrounded by his enemy with only a few old solders, he chose to play music on top of the city wall with wide-open gates. His enemy got suspicious. They could not believe that he had a weak army inside. They suppose that there must be a strong trap. They retreated, and Zhu won crucial time to get support of the large army elsewhere. This tactic has been used in different ways. The principle is not to show your weakness. Blowing up oneself when one is weak is one way; emphasizing strong points to hide the weakness is another way. Many young Chinese also use this tactic a lot to get a good position by a foreign company. Because of their much better English, they are also most of the time favored by the foreign recruiters.

A utility company wanted to set up their office in Shanghai. They have developed special coal-based chemical waste treatment technique to turn most of the waste into useful resources. It is a very specific product for a niche market. To get to the market, they need to have the right network. After being advised on the existing Chinese network and the market access, they rightly preferred a Chinese manager. This manager needs to communicate with the headquarters and carry out the strategy. From then on, difficulties started. The consequent continuation and implementation of the China strategy would be to use "Chinese" judgment and selection procedures to get the best person for the job. But they would like to have the Western way, so a Chinese-based headhunter was hired. They got hundreds of candidates with excellent CVs. To select the good one, they used the price criteria. In the Western market economy, when something is rare, the price will be high. Their reason to interview the ones who require higher salaries is based on the reverse of this principle. They thought that the one who asked for higher salaries must be more wanted. They chose to interview the expensive Chinese. Most of them are not "real" Chinese but overseas Chinese like American, Singaporean, Taiwanese, and Hongkongnese who are used to have higher salaries. A young Hongkongnese who studied in the USA and had been working for a very impressive list of multinationals was chosen. In his 8 years of work experience, he was appointed nine times as sales or project manager. He had carried out more than eighty environmental projects worldwide. In the eyes of the client and the headhunter, he gave a very good impression in the interviews. When being asked for references, a few big names in the international NGO world were dropped. In Western eyes, he seemed to be the ideal candidate, but Chinese eyes would have arranged additional intelligence about his background. But the utility company wanted to start quickly, and the decision was taken to hire the candidate. He left the company 2 months later for another company which was willing to pay more. Then our client understood where his long list of employers came from. Later on, it was found out that his list of projects was also not what he carried out but what his former employers had carried out. The famous people he knew were the people he saw at a conference.

Both Chinese and European companies are mostly careful with people who change their jobs very frequently. Sometimes, such flyovers can transform company and hand over the newly created organization to going concern managers. Sometimes, they are luck seekers without added value for an organization. Therefore, key is how to deal with it. A Chinese consultancy firm had in the eyes of Westerners a very unusual way of selecting people. One of the first things they ask in the job interviews is about the family. "Are the grandparents still alive?" "Where does the grandparent live if only one is alive?" "How often does the candidate visit his parents if they are not living together?" and other questions like these are the questions that matters. If the employer finds out that the Chinese candidate's grandparent were not taken good care of by his parents or he does not have warm bands with his own parents, the candidate will not be hired.

4.4 Business Model

4.4.1 The Wrong Strategy

The camel bird with its head in the sand. An international service company organized a client seminar for their Austrian branch consultant organization about the upcoming Chinese economy. It was planned to start a cooperation to develop China business services for Austrian companies. One of the starting points is that they need to have a China proof strategy. Some CEOs of local Austrian companies approached us and told us very proudly that they do not believe in strategy, let alone China strategy. Many Austrian companies in the branch are middle and small business with innovative technologies. Some of them are world-class leaders in their subindustry but unknown by the general public. They either do not have the money or the interest in developing a good strategy. Because of the attractiveness of their technologies, they were very welcome in China at that time. The huge interest and potential demand gave the wrong impression that a strategy was a waste of time and resources. More than a half decade later, the interest in China business services is increasing. The gamble strategy and world-class image did not work, and bad experiences slowly make clear that business development in China despite excellent technology needs something more. For example, a start-up approached us for investment from China because the source dried up from European investors; the technology owners are fighting the growing Chinese competitors; the suppliers need to go to China because their clients went to China; the service providers could not follow their clients to China because of lack of local partner; and "made in Austria" wants to develop the Chinese market because of the declining European and American market and rising value of RMB.

When the name is wrong, the cause will be wrong: This is an old Chinese phrase about finding the right definition of the strategy. The most common mistake a Western company makes is to go blindly to China without having a clearly view of their strategy or business development plan. Even for those who take a step-by-step approach and are aware of the pitfalls, it remains very difficult to deal with the Chinese on the long term. Many of these organizations either have enough of the not working strategies they have got made by expensive consultants or have too little resources to finance the preparation for China. Not having a China strategy is often fatal for an organization working with China. A bad one is even worse. But a good strategy on paper is not always enough. A political leader of a prosperous region in Western Europe prepared his first visit to a Chinese province as the leader of the business delegation. Under the impression of the Chinese economic development and the enthusiastic reactions of the trade delegation, he realized that the companies from his region need the governmental support. He asked us to develop their China strategy because we have experience working with Chinese government and we have high-level and long-term contacts with a few of their potential partners. Another company had organized a conference for the province with a city in China. They also wanted to develop the China strategy for the same province although their core business is organizing conference. From our point of view, there should be support and contribution from all stakeholders to develop a workable and fruitful strategy. Our special attention went to the process of developing the China strategy and getting it realized though our network and striving for bottom-line results. Both a top-down and bottom-up direction communication line was established between the involved parties. The existing networks should be optimally used because Chinese trust old friends better. The stakeholders including the other company which was actually our concurrent are invited to give their opinions. Unfortunately, one of the advisers involved used the provided information to contact for their own interest and their own connections in the province. Both advisers we asked to develop an implementation of our strategy. Our implementation strategy is simple and effective and boils down to a pyramid of the leadership friendship as tracker, the civil service cooperation as the pillars, and the company cooperation as the base and target. The alternative strategy was based on big names, nice fancy organizations, information platforms, and impressive website. The latter one won. Forms and design wins from content and result. But the form and design implementation looked nice on paper but did not work in practice. Our network in the Chinese provinces refuses to work with them because of their loyalty us and their trust in the results of the content and result approach fitting to Chinese pragmatism. They could only work with the Chinese city they knew. The nice organization on paper turned out to be an empty shell because they do not have capable people who understand both sides at strategic level; the companies which wanted to reach China through the government are mostly disappointed because the level of the contacts is too low. This case is typical for governmental organizations which have to defend their plans based on paper promises and much less on concrete results. Chinese organizations also have paper plans full of promises, but frequently, they also want to achieve pragmatic results because their business partners in the guanxi network expect that from them. In return, when successful, the government gains power which is often a major interest to further develop their territory.

4.4.2 The Penny Wiser

Many Western managers in China work with bottom lines either from the headquarters or from themselves to make career. Long-term investment is not always the priority as the next case from a big company shows. They want to sell certain environment technology to the Chinese. After the market research, it became clear that the biggest potential client of the technology would be the government. But there are several ministries involved in this case. The next step was to find out the key players and the best way to approach them. One of our Chinese partners has the contacts. As the semigovernmental organization under one of the ministries, they are a kind of house consultants. The boss is one of study mates of the coauthor graduated from environmental sciences department. Because almost every ministry has a kind of environmental department, our study mates can be found in every ministry. The Chinese manager knew that he needs support from different sides he had made friends within the other ministries long before he needed them. Because of these useful connections, they have been carrying out assignments for the ministries for several years. We discussed this possible partner with the client. The environmental company was very interested because of the existing network. Because we knew the Chinese boss, it is also easier for them to be introduced. In the beginning, the Dutch was very happy with the new partner. The Chinese were very cooperative, and they could get some assignments together. But soon, they found out that they have different ways of getting assignments. For example, the Chinese company used to organize free conferences and training sessions in beautiful places for the people from the ministries who are in charge of the case. This is very unusual for the Dutch company. When they give training, they always charge for it. The Chinese way raises the cost of course. This gave a lot of frictions on both sides. The last straw was when they had the chance to win a tender by one of the ministries. The tender was about getting access to a countrywide market. The Chinese partner insisted on having a conference in the home country of the company, and they should invite representatives from all the different ministries to demonstrate the technology. The company leaders were reluctant because of the cost. Only after some firm discussion, they were persuaded to do it anyway in the sake of future cooperation. To save money, the Dutch organized a one-day conference without any side events. When the Chinese guests were in European headquarters, the Chinese boss was very nervous. He required several times that the partner should also take the Chinese guest to sightseeing trips. But it was weekend after the conference. The top managers chose to stay with their family. They did not get the tender. The old friends were not friendly to the Chinese boss any more. He had to choose between the old Chinese friends and the new friends. He chose the old ones. The cooperation ended.

4.4.3 The Ghost Market

In the last decades, attracted by work or a better life in the big cities, many immigrations from the countryside flood the Chinese cities. This makes the already huge housing problem in the big cities even worse. Towers of apartments were built every year. In all, the first-line and second-line cities, despite the new apartment towers like one tune concrete forest in the cities, the price of housing has rocketed by scarcity and speculation for years until 2008. This trend was picked up by an eco-housing architect. In 2008, troubled by the declining market because of the ongoing real estate crisis at home, they wanted to have a try in the booming Chinese market. They visited some big cities following their Minister of Environment in the business delegation. Because of the presence of the minister, some Chinese local government leaders and big real estate developers were also present. Almost everywhere, the concept of eco-housing was welcomed. Soon after the trip, they got a small assignment to design a model house with zero emission for one of the real estate developers. The project was also sponsored by the local government because they have very large green energy industry in their region. A mega project after the tryout is proposed by the local government. They even showed the piece of land to the architect where the eco-houses would arise. Money for the mega project was promised by several interesting Chinese and foreign investors. All signs about a booming market are showing. The architect finished the tryout project. But the mega project did not come. The housing market in the region went slow because of the changed mortgage policy. The Chinese real estate developer was bankrupt because of the unsold apartment towers they built. The local government could not provide the land because of the changing land use policy. At the same time, Chinese architects discovered the eco-housing potentials and learned to design with the concept. The booming market for the foreign architect turned out to be a ghost market. The only thing they can do is to team up with a Chinese architect bureau.

4.5 Institutionalization of Innovation

4.5.1 The Valve of the Pressure Cooker

A big Chinese electronic company was worried that they cannot catch up with their foreign competitors. Most of the money they earned was by sourcing for the foreign companies. They need to develop their own products for the home market. The company was run by an old military who acts like a general too in the factory. In the years they sourced for the others, quality was sometimes a problem. He showed his temper a few times when the responsible ones were found out. The quality went up. But the people working for him were all very afraid of him. When he first launched the plan for innovation, nobody dares to come with an idea. He got an idea. He hangs up a black box outside his office with the name "idea box" on it. Everyone is welcome to put his idea anonymously in the box. He promised that all ideas will be read and the good ones will be announced. The good idea owner will be invited to come forward and take charge of the development project. When it turns out to be a good idea which brings money for the company, the owner of the idea will get shares of the company. The idea box has been introduced in the West many years ago and in many cases has been abandoned or replaced by a more structured bottom-up process. For Chinese companies, organizing the bottom up is a far bigger step than for most Western companies. The idea box is a great step, but in China, it will often remain empty because it is difficult for Chinese to put forward their own individual ideas without knowing the bigger context.

4.5.2 Behavior of a Gambler

Chinese are not rebels most of the time. But when they have the chance to explore or when they are forced to take chances, they can be very good innovators. They usually act like a gambler. Many of them will not make money. But the few lucky ones may hit the jackpot.

Mr. Zhang is a so-called sea turtle. In Chinese, "sea turtle" sounds the same as "the one who returned from abroad." He was once a civil servant in the 1980s. Many Chinese would envy his position having the "iron" rice bowl. He was also very glad with his work until his boss approached him with the proposal of introducing his daughter to him. The daughter was much less pretty than the girlfriend he already had. Mr. Zhang took his chance and chose for love. Soon he found out that his working life turned miserable and that was not only because of bad luck or coincidence. He had to find a way out. His parents had a far relative in the USA. He went to America to do Ph.D. in physics. His parents and pretty wife lost their jobs at home, but he was too poor to support them. So he started a small business in the campus in developing Chinese software. Several years later, his company was bought by a big Silicon Valley star. He stayed in the new company and learned how the American way of innovation generates new ideas and new business. In 2000, the Chinese government realized that China was brain drained and started a campaign to attract those talented Chinese back to China. If they want to work, they can get high-level jobs in the government or state-owned companies. If they want to start their own business, special founds will finance the business. Mr. Zhang chose to use his knowledge of generating innovation in China. He started an incubator in his hometown with some of his own money and some local government funding. He recruited some of his Chinese friends in the USA to develop the business. But he soon found out that the people who are very innovative in the USA are not as innovative in China. The China he missed so much is not the one he can develop his business. A few promising business like producing LED lamp went bankrupt last year. After losing a lot of money, he had to change his business strategy. He joined forces with incubators in America. He now owns a multinational and multi-sectoral business from electronic, health care, to agro-food. All those will be crucial according the big trends. He still wants to take the chance to hit the jackpot with one of the promising sectors.

4.6 The Big-Picture Approach

Chinese environmental engineering sector was not famous for its advanced technologies. But it has seen a growth which its foreign competitors with much better technologies will envy. The growth is not always due to the stimulation the

Chinese government gives. In most of the cases of a private company, it is because they knew their end goal. From left or right, while looking for the missing puzzle to reach the end goal, they unconsciously chose the right way to innovate their business. We summarize it as the big-picture approach.

A Chinese professor did research about how to treat certain wastewater. His research was sponsored by a pharmacy company. He knew if a company wanted to spend money in something which is not their core business, the company must be in need. He let one of his students find out the market and the potential market due to the stricter standard in the years to come. It turned out to be very promising. With this big picture in his mind, he knew money can be made with his knowledge. First, he started a research institute in his university. After not being treated well, while he collected necessary wastewater treatment technologies, he set up a consultancy company of his own. He soon found out that his knowledge was used by engineer companies to build wastewater treatment installations while he was being paid badly or not at all. When he wanted to advice pharmacy companies using his network, he was always welcome but never got a project. The companies asked him not only to give advice but to solve their problem. He used his consultancy company to form a joint venture engineering company with a construction company. Then he could offer a turnkey project which is preferred by the pharmacy companies. To get some clients, he formed a board of advisors with retired officials who still have network in the right government departments or former top managers in the pharmacy sector. To control the advantage of their technology, he traced all his former students and got the most of them back to work for him with promise of positions or shares of the company. It is now a matter of time that his company will drive the other engineering companies out of the market.

4.7 Governance

4.7.1 The Invisible Power of People in China

"Water can carry a boat; it can also turn a boat." This phrase is used to describe the power of ordinary people in China. Most of the time, the Chinese folk are very tolerant. The Chinese word of "tolerance" contains a "knife" above a "heart." From childhood on, we learned to obey the authorities and to harmony with others, especially the one with more power. But this does not mean that the tolerance is endless. When a powerless Chinese is forced to a corner with no way out, they can be the worst nightmare. Chinese people know their rights. They also know how to organize it when it is necessary. "One chopstick is easy to break, a branch of chopsticks is not easy to break."

Many Westerners do not understand this. They are always surprised that the mostly kind Chinese they coped with can turn their back to them. A chemical company set up a factor in China in the late 1980s. At that time, there was big shortage of fertilizer and pesticides because of the booming of intensive agriculture when the Chinese agriculture reform policy was carried out. The company also chose China because of the milder environmental policy in China at that time. The local government promised that they will be left along if they stay inside the environmental law. The standard was much lower in China than their home country. That saved the factory a lot of money not to clear up the pollution. For years, they could discharge wastewater containing poisonous chemicals. The local government was always very friendly because of the tax they generated and the status of having a world famous company in the region. The people were friendly too because in most families, there is someone working in the factory or in the supplier of the factory. In the last few years, the rate of cancer in the region has raised to six times the normal level. The region is now much richer. People give more concern for their living surroundings. After hiring a university outside the region to test the water, they soon found out what might be the source of the unhealthy environment. The people were not friendly anymore. First, only the families of the cancer patients protest outside the factory. But the local government was reluctant to enforce the now much stricter environmental standard. After half a year without any result, the family moved to camp before the local government building. A few was arrested because they tried to disturb the public order by holding on to the legs of some officials going to their work. Then the people were furious; thousands of Chinese tweets were posted. All villages drinking the polluted water were informed. Large-scale protests were organized. Reporters were invited to film the helpless crying family in the angry crowd. Other than the Westerners, Chinese leaders know very well when the people's opinion should be respected. When the news reaches the regional leaders, they could not stay calm anymore. The factory was ordered to close immediately. The jailed famers were released without being charged. The foreign managers had to flee because their homes were surrounded by angry Chinese. The Chinese production manager was arrested and trialed in public. Besides the financial loss, the company will never be able to recover from the image damage they suffered in China.

4.7.2 The Tiger with Wings

How can a tiger fly? To have some wings. Strategic partnering is a very common tactic in the Chinese environment sector. The tigers are mostly big companies. The wings are usually government-related organizations. They are not always state-owned companies. Some organizations are even very small in the Western eyes. Because the role of the government in the environmental sector is still very big, the favors of the government are always a concern of a Chinese company in the sector.

There was once a poor region in central east China. In the history, it was once famous by the fish and rice from the region. The last 30 years, the river was polluted, and less and less fish was caught; the land was over cultivated and provides poorer harvest each year. Seeing other Chinese regions getting rich by stimulating policies from central or local government, the local government also wanted a new economic impulse. They found out that the poor land they have contained a hidden treasure: sand, in fact high-quality sand which is needed for the upcoming solar cell industry. A solar investment agency was founded with only a few people working for it. The agency started organizing investment events to attract investors in 2006. A smart rich Chinese approached the local government. He promised to invest in developing the solar cell industry for the region. He had enough money to start the silicon sand factory. The only requirement he had was that the local government should be the shareholder too using the agency as vehicle. The government did not have much money to invest, so they gave the exclusive right of using the land and to explore the mines. Because of the booming of the solar industry the last years, their high-quality sand was well sold. The company went public, and the shares rocket sky high. The government also got rich with their shares rising. The government decided that they could do better. So they decided to use the money they collect in the stock market to invest more in the industry. All new buildings and governmental facilities have to use solar energy. The local households can buy the solar panels with cost price. Many people use this advantage to start their small-scale solar plants in remote areas without electricity like Tibet. It generates more revenues for both the company and the government. The region is now enjoying prosperity. They have now built up a solar imperium providing the whole chain of products and services. This year, when most of their rivals are facing financial difficulties, they still can stay standing by pumping government money back to the company. This company really is a very good tiger with strong wings.

But one has to remember that the wings on a tiger are not a natural thing. Sooner or later, the tiger has to realize that he cannot fly all the time because he may lose the wing. The coauthor of this book finished her Ph.D. in 1996 and went to work for a Chinese management consultancy company established by the Chinese central government. The founder of the company is a Chinese who studied management system in England. He knew that there is a shortage of good management services for most of the Chinese companies at that time. After successfully building up the ISO 9000 services, a new department is set up to serve the clients wanting to have ISO 14000 certificates. Thanks to her Ph.D. title and some network in the Chinese environmental world, she was asked to be responsible for building up the ISO 14000-related services. One of the first things we did was to team up with government. We trained all the related officials. We made the Chinese version of the standard together with the Chinese central government. We got the permits to select Chinese blue chip companies to build up their environmental management system at pilot projects. When she left for the Netherlands, a few years later, the company was doing very well as one of the biggest consultancy company in the field with tripled profits every year. But it did not last long, because many semigovernmental organizations like research institutes also discovered the market. They are traditionally twinning with the governmental all the time. Soon, hundreds of different companies were generated by those institutes and offering the same services to the local clients. Supported by their local governmental partners, they also got most of the clients. The company still exists but is forced to operate with much less financial profits as the first few years.

4.7.3 The Fox Behind the Lion

A Chinese children's story is about a smart fox who knew how to use the power of the lion to be the king of the forest. The fox told the lion that he was the king of the forest and all animals were afraid of him. So the lion should also be afraid of him and gave him presents. The lion asked for evidence. The fox answered very proudly: "Come and walk with me through the forest, you will see what I mean." Everywhere, when they appeared together, the other animals ran away frightened. So the lion believed the fox and brought meat to the fox afterwards. The story ended with the lion found out the true power of himself and killed the fox.

Despite the moral of this story, this is still a well-used strategy of Chinese companies to get Western technology. Most of the Western "lions" still do not realize that they are misused and still bring more "meat" to satisfy their Chinese partners. A Frenchman used to be a top manager of a multinational. He worked in Hong Kong and China for many years in the 1990s. At that time, he made some "powerful" Chinese friends. These friends used to take him to their hometown (mostly remote areas) to develop the business. Everywhere they went, they were received like kings. In his eyes, it was due to the "connections" of the Chinese. But in fact, the real reason was that he was the big boss of a world famous company which has state-of-the-art technology. All the local government would like to have the business in their region. He was convinced by the "power" of the Chinese friends. Several joint ventures were set up in different regions where the Chinese friends came from. The friends were appointed important positions in the joint ventures. Ten years later, only a few were still his friends. Most of them turned into his rivals having their own companies in their hometowns making almost the same kind of products. The French company tried to fight in court for the IP rights. They lost in most of courts because the Chinese were using much better technology which was a combination of the French and the Japanese one.

A famous multinational in the green business wanted to enter the Chinese market. After studying the market, they decided to go the easy way. Acquisitions were already a proven strategy for this company. So buying a Chinese company seemed like a right thing to do. But which one should they buy? They hired a successful business bank to guide them through this process. This business bank is very successful in China because they have a Chinese top manager in the Chinese branch who has very powerful background. A selection of companies was made. At the same time, the Chinese top manager left the business bank to start his own M&A business. He contacted the multinational and wanted to take this business with him. The multinational did not do it. But they could not go on talking to any of the Chinese companies selected by the Chinese top manager. They bought another Chinese company helped by the business bank. But their new Chinese adventure did not last long. They are soon pushed out of the market by the ones they could not buy. In this case, the business bank is the fox behind the powerful Chinese top manager.

4.7.4 Law and Order: One Eye Open, One Eye Closed

Alongside of a beautiful river lies a city in Northeast China. It used to be a poor region. Several years ago, gold was discovered. To stimulate the local economy, the mayor decided that they should attract mining companies to the region. As a good Chinese leader, he took quick actions. Meetings with the municipal departments were organized to make sure that they developed quick and total services for the mining companies. One-stop service was born. The mining company could at once get all the permits including the environmental permits to start their business. The environmental permits required that the total quantity of pollutants should be under control and with the environmental standards. Mining activities soon brought prosperity to the place. But they also brought serious pollution problem. The local Environmental Protection Department found out that several mining companies were not treating the wastewater according to the environmental permit. Some even did not have an environment permit. They warned the companies by sending them official requirements and threat to fine them. Calls from the mayor's office made it very clear that the environment should not be the priority. The advice was to keep one eye open and one eye closed. This prevented the Environment Department to take any actions. After a serious environment incident by a mining company, people got to know the price of the economic benefit. Luckily, there is still one eye open by the Environment Department. They could take very quick actions once the mayor saw the necessity. They picked out a few big mining companies and gave them very heavy fine. Soon, the polluting activities were stopped because the small ones went away or bankrupt, and the big ones were afraid of the heavy fines.

4.7.5 How Water Drops Can Penetrate Stone

A world famous environment NGO started their activity in China 30 years ago. Chinese NGOs like women's associations or labor unions were actually under governmental supervision. They had much less influence than NGOs in other countries. At that time, China was not as open as now. It was also very sensitive of the critics from international communities. This NGO did not choose the finger-pointing policy. They knew that the Chinese government was powerful. They had to get the Chinese government on their side if they wanted to achieve anything. Step by step, they first improved the awareness by organizing trainings and study trips to other countries for the officials who are responsible for legislation development; pilot projects were selected together with the Chinese; a lot was achieved by working with the central and local government. Nowadays, they are welcomed guests in forums to discuss the possibilities of improving the environment legislation system in China. Long-term vision, patience, and persistence—this is the way how a powerless NGO gets into the center of power.

4.7.6 Dragon Head Tactic

"Dragon head" is a Chinese term used to describe famous organizations in their branch. If the Chinese government wants to carry out a policy, very often, a dragon head is pushed forward to be the showcase for others. In the case of implementing the environmental law, dragon head tactic is frequently used too. This tactic can be used both ways, mostly in the positive way to show how dragon head can take the whole dragon to fly. But sometimes, especially in critical cases, dragon head can be cut, too.

In 1996, ISO 14000 was introduced to China. Working for the first consultancy company in this field, the co-author got to know a lot of dragon heads. A list of best companies in different branches was made together with at that time State Environmental Protection Agency (nowadays Ministry of Environment). Before the standard was converted into Chinese National Standard (GuoBiao), those dragon heads were encouraged to implemented ISO 14000 in their companies as test cases. After a year, experiences were learned, and not practical points were discovered. Those feedbacks were taken into consideration when the National Standard was published 1 year later. The companies which participated in the pilot program wanted to show off their environment protection achievements. They marketed the results of their ISO 14000 certifications where they could. Because the dragon heads were very famous organizations, ISO 14000 was rolled out in China without a lot of publicity efforts.

4.8 Lessons Learned

Only already this limited number of cases contains a few lessons for how to do business in China and to grasp the opportunities of Green China. Just as with dos and don'ts, these lessons only give you hints, and it is only the top of a big iceberg. Sometimes, depending on the time and occasion, the lessons do not even apply.

Lesson 1: Right leadership is the key to success: In China, the modern Western ways of leadership such as being supportive instead of decisive are not very common yet. A Chinese leader is a real boss implying taking the big decisions and being carried out by subordinates. But we have seen that sometimes a wrong decision can be taken in which the flexibility and fast-moving decision making can be a way out.

Lesson 2: Business tactics: Chinese leaders make use of a wide range of available tactics which most Western leaders are not aware. Improving the knowledge of China business tactics can help Western company improve their performance in China.

Lessson 3: Being yourself: Being yourself is authentic but is not a sufficient condition for success. Being yourself helps you to be trustworthy, but still adaptation to ways of doing business if your markets or business partners is required.

Lesson 4: Making money: The Chinese market can be a very profitable market if you know yourself, the real markets, and their relevant players are known.
Lesson 5: Controlling the Chinese: As soon as you think you control the Chinese, they control you. Or they try to do so in the future.
Lesson 6: Arrogance and success: Having seen it all and knowing your business and thinking you can do it on your own often turn out to be loss strategy. Success in the short term can turn into a loss giving long-term situation.
Lesson 7: Winning alliances: The winners are not always the ones with best technologies but the ones with best connections and the acknowledgement of the interests of the partners. Chinese prefer to do business with other Chinese they know. The dynamics of their connection determines the outcome of the alliance, not the Western contribution and input.
Lesson 8: Innovating business model: A Western company needs a suitable and flexible business model in China. The business model should at least pay attention to the following combination of contents: long-term vision, leadership, adequate Chinese team, good HRM methods, feeding and using Chinese network, right marketing, workable procedures, and good timing.

References

Someren, T. C. R. van, & van Someren-Wang, S. (2009). Opgeheven wijsvinger of opgestroopte mouwen? Milieu, 28–29.
Someren, T. C. R. van, & van Someren-Wang, S. (2007). Strategic eco-innovations, proceedings 1st water conference province of Hebei, China, June 2007.
Someren, T. C. R. van, & van Someren-Wang, S. (2007, March). "Innoveren in China": China wil meer zijn dan fabriek van de wereld. *SER Bulletin*, *3*, 14–17.
Someren, T. C. R. van (2006, April). Strategic innovation: A matter of leadership. *Business Digest*, *162*, 9–10.
Someren, T. C. R. van, & van Someren-Wang, S. (2007, February). What style of leadership is required to succeed in China? *Business Digest*, *171*, 1–4.
van Someren, T. C. R., & van Someren-Wang, S. (2007, July 23). Five tips for dealing with Chinese. *Science Business*, on line publication.

5 Strategic Innovations Create a New World Order

On Strategy and Innovation: The Worldwide New Fundaments of the Future Economy

The rise of China accompanied with grand plans of the Chinese central government, new industries and firms, the different way of doing business, and creating new world order with new rules of engagement forces the West to rethink their policies and business strategies. The time of simply following a low-cost strategy by outsourcing or a market development plan is over. Strategic innovations have to be created of which creating tailor-made business models is merely one aspect. This holds not only for the USA or Europe but for China as well. The Strategic Innovation Theory (SIT) offers the general approach to create your own necessary renewal to cope with Green China. In particular, the Strategic Innovation Theory can be specifically applied on the issues of sustainability. The Ideal Sustainable Development Corporation is a special strategic innovation which fits to the issue of Green China.

5.1 The Birth of a New Market Universe

Green China and the response of the EU and the USA need strategic innovations and not only new business models. Why? Because business models are related to the earning capacity of individual organizations. Greening of China also involves central and local governments, the differences in Chinese and Western leadership, cultural differences, the ability to institutionalize innovations, and many more. Together they are called strategic innovations which go beyond business models (van Someren 1990, 1991b, 2005).

Sustainability in general and Green China in particular are dependent on the ability to create and implement strategic innovations. But also the reaction of Europe and the USA on Green China is dependent on strategic innovations. In this chapter, we give a short introduction of the background of strategic innovation

T.C.R. van Someren and S. van Someren-Wang, *Green China*,
Management for Professionals, DOI 10.1007/978-3-642-28810-4_5,

and its importance in relation to Green China. The basis is Strategic Innovation Theory (SIT). We will use some core issues of the SIT to understand the development toward Green China and to be able to give a Western answer[1]. These core issues are the relevance of nontechnical innovations, new growth curves, new business models, and context. When taken together, these issues form strategic innovations. Strategic innovations create a new universe where different market games and rules will be created and applied. For example, the organizational principles of Just-In-Time led to new factory designs and new cost–benefit calculations, thereby revolutionizing the way of manufacturing.

5.2 Nontechnical Innovations

Innovation is not technology. Nontechnological innovations are equally relevant as technology for growth, development, prosperity, and the way we do business (van Someren 1990, 1991b). Among the nontechnical belong elements like new designs, new organizational forms, new products and services, new performance criteria, new value propositions, and many others. However, during time innovation is being interpreted as technical innovations. Especially, policy makers and academicians favor technical innovations because they are concrete and often physical present. You can touch, feel, taste, and smell it. It is sexy. The consequence is that the category of technical innovations has got almost a divine character and a solution for every problem. This can often be seen at policy makers, academicians, and technical firms. For them, it is a mathematic equation: Innovation = Technology. The contrary is true. Nontechnical innovation can be as important as technology and emerge stand alone. Moreover, nontechnical innovations can lead to technical innovations. Figure 5.1 gives an overview of nontechnical innovation areas.

Nontechnical innovations encompass organizational forms, value chain organization, foundations of behavior and performance, new design, ways of doing business leading to new ways of doing business like the introduction of Triple P instead of profit maximization, network organization, and outsourcing. Sometimes a combination of nontechnical innovations leads not only to new technical innovations but to new business models as well (van Someren 1991a, b, 2005). An example of an organizational innovation leading to a bunch of new technologies is the integration of the R&D laboratory within the organizational boundaries of the own firm. Illustrations of combined nontechnical innovations leading to new business models are franchising, supermarket, and Just-In-Time.

With respect to Green China, nontechnical aspects of strategic innovation have several faces. It can be related to the closing of material cycles requiring besides technical solutions the physical–economical–organizational connection between

[1] A much more detailed elaboration of the Strategic Innovation Theory and basic background literature on theory and practical appliance in particular firms and policies can be found van Someren (1991, 1995, 1998a,b, 2005).

Fig. 5.1 Strategic innovation areas (examples). (Source: van Someren 1991 and 2005, Strategische Innovationen, Gabler Verlag)

disposal of products and processes and resource input. Another example is the relevance of stakeholders in determining which elements of the Triple P are regarded to be relevant.

5.3 Jump the Curves

What is the relevance of technical and nontechnical innovations? These kinds of innovations can help to improve existing business. In fact, the lifetime of the existing business model is being extended. For example, with new IT systems, existing customer relation processes can be made more efficient. Or with cross cultural management, training can improve the relation between headquarters and foreign affiliates and business partners. The majority of firms, smaller and bigger ones, behave in this way. But there can be gained more than only improvements. The real relevance of strategic innovation is the potential of creating a new growth curve generating new business. In this perspective, strategic innovation is in fact the

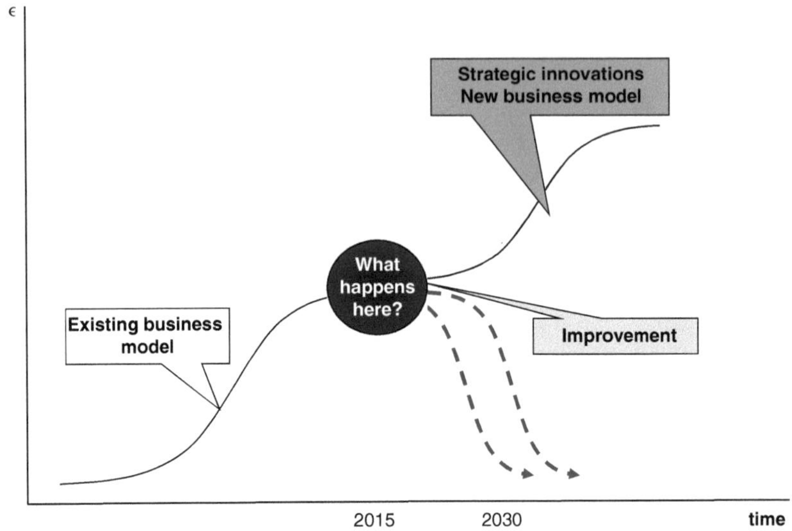

Fig. 5.2 Relevance of strategic innovation (van Someren 2005, Strategische Innovationen, Gabler Verlag)

central task of entrepreneurs and top management. It shows the pivotal role of strategic innovation for any business.

Single innovations relate to new products or new services, new organizational forms or a new financial instrument, and many other areas. Strategic innovation refers to a unique combination of innovations leading to a completely new way of doing business. A new business model has the potential to revolutionize existing value chains or creating new ones. In Fig. 5.2, this potential effect is illustrated.

Of course any entrepreneur or manager immediately recognizes the potential of a new growth curve. But two main questions will pop up. The first is how can we identify where we are on these curves. The second one is if we want to jump from the old curve to the new one, do we know what happens there and what we should pay attention to.

We turn to the first question. It is sometimes indeed not easy to identify on which point an organization, industry, or even nation is. But when looking carefully, weak and strong signals can be collected. Strong and especially weak signals in society and individual organizations indicate the potential of creating a new wave. In Table 5.1, a few illustrations are presented.

To reduce complexity, we make a difference between macro, meso, and micro levels.

The macro level is the level of the world economy, continents or economic regions, or macroeconomic indicators like gross domestic product (GDP) or invested R&D as percentage of R&D. On a macro scale, in the old growth perspective curve, countries can have a focus on inland problems, thereby overseeing relevant developments outside their own borders. Another indicator is

Table 5.1 Features of old growth curve versus new growth curve

	Old growth curve	New growth curve
Macro	Inward-looking countries	Outward looking
	Lack of new visions	Progress on basis of future dream
	Focus on minor en detailed problems leaving aside big world issues	New vision of the world
	Stakeholders of minority positions dominate nation's development	Proactive policies
		Stakeholders of the big new future determine decision making
Meso	Defending old industries	Development and growth of new industries
	Lobbying dominate policies	Entrepreneurial activity
	Policies and discussions without action and implementation	Action planning
Micro	Decreasing profits	Competition among rivals for new growth
	Low-revenue growth	Innovative growth steps
	Improvement of existing business model	
	Defensive	

the attention of management to minor problems instead of contributing to larger world-scale problems. Discussions in several European countries on raising social security payments by a few euros a month and not being capable of saving the euro crisis illustrate this indicator.

Matured nations have organized their nations in a complex accumulated system of law, regulation, and institutions that govern the country. Interest groups influence politics and institutions trying to get interest in minor problems, leading to improvement of the existing nation system. Laws and regulation help existing stakeholders and not the renewing forces. Examples are discussions about a few euro per month more or less for social security receivers and long discussions about improving infrastructures like roads and railways. The end of BBC World Service broadcasting in 2011 is a sign of retreating world power. The appointment of a Chinese member of the IMF Board is a sign of increasing power of emerging economies. The knock on the door of the EU at the BRIC countries for financial support to save the euro will not be for free. BRIC countries will ask a price.

These are all small but relevant shifts in power, and the accumulated effect is the development of a new growth curve. Budget cuts in European defense sector show the financial constraints. Discussions whether or not immigrants should learn the language of their new home country are going on for about twenty years. This is not stakeholder bottom-up democratic decision taking; this is a swamp where only the Baron of Munchausen can find rescue. Valuable time is spent on minor issues, thereby missing the relevant issues of the next growth curve.

The meso level refers to the level of industries or regions. Protecting large companies in old industries like oil business extends the lifetime of the existing growth curve. These large companies have the resources for lobbying and influencing regulations in their favor. The new growth curve is characterized by entrepreneurial activity in new directions like renewable energies. Also policies to stimulate new ventures show the commitment to old or new activities. Big new infrastructural projects are announced as the rise of the nation and job opportunities for the people. New industries and companies arise accompanied by extremely rich entrepreneurs; inland migration for job seekers contributes to a flexible community. As Green China shows, economic progress is top priority and detrimental environmental effects are a calculated loss to be tackled later on, and country's interest is bigger than individual's interest. Opportunities win from risks and threats.

The signs on the micro level are revenues and profit margins' decline and rising cost levels. Improvements are on the short-term cheaper than doing something radically new. Innovative projects are rejected because they do not fit to internal criteria of projects. In this context, new industries are getting attention from scholars and policy makers and entrepreneurs, but because of fundamentally different features, it does often not fit into established society.

In Europe, contrary to the USA or China, genetically modified food is labeled as Frankenstein food which disables a fundamental discussion about pro and contra. Incumbent energy companies grown up by oil, gas, and coal-fired plants stick to their existing business model and have no eye and feeling for renewables. Time is lost due to discussions about what, how, and where. In the scenarios for 2030, the big oil companies all believe that fossil fuels will remain dominant in energy supply. BP estimates that by 2030, the share of all fossil fuels (oil, coal, and gas) in total primary energy will be 26–28 % each and of nonfossil (hydro, nuclear, and renewable) around 6–7 % each (BP Energy Outlook 2030).

The reason is not, as the big oil companies say, because of better technologies or better price per energy unit but due to improving the existent business models based on existing upstream and downstream infrastructure such as geological knowledge, drilling techniques, pipelines, and distribution. The decreasing share of oil is replaced by gas supplies. Only bio-based fuels are an option for the incumbents because they fit into existing system of refineries, transport, and distribution. Both BP, euphemistically Beyond Petroleum, and Shell made some attempts to invest in renewable energy. Shell started with solar PV, but discussions about wrong accounting of their proven reserves lead to a full focus on oil and gas again. Markets were afraid Shell would not survive if reserves would be too low, and exploration for oil and gas became priority number one.

It shows that most of the time the incumbents, although they have all the resources needed, are doing business in their own prison of existing business. In their Energy Outlook, BP observes that the energy intensity measured as toe (tonne of oil equivalent) per thousand $2010 GDP is decreasing worldwide, converging toward 2030. As a result, economic growth will be less intensive. The message is, in the long run, BP and other fossil fuel companies are able to provide the world of a different mix of fossil fuels, and the big oil companies remain inevitable for energy supplies. This would be right if we stick to their strategic views and leave aside the opportunities of strategic innovation (van Someren and van Someren-Wang 2010a). But the BP Group chief economist, Christof Rühl, defends the old business model with referring to mechanisms of a mature growth cycle. According to Rühl, the reasons behind the decreasing energy intensity are the energy trade, the diffusion of technology, and the standardization of consumption baskets (BP Energy Outlook 2030, p. 19).

Moreover, individual organizations often show an inward-looking vision focused on defending existing interests and improving current business models. Innovative projects are difficult to realize due to a complex web of regulations stressing the impossibilities instead of the opportunities. No decision is being made due to stakeholder interests and their influence on decision process. Contrary views exist in upcoming countries and organizations.

These indicators can be applied on both private and public organizations. For private organizations, the indicators like profit margins are rather straightforward. Public organizations lack these relatively simple measures, but even then alternatives are available. As soon as the indicators lead to the conclusion that continuing doing business on the old-fashioned way should be questioned and replaced by something new, the time has come to rethink and question your own business. Now the second question comes into the picture.

Answering the second boils down to what top management thinks about the future. Nobody can exactly predict the future, but it is possible to create your own future. When top management concludes that it is time to jump the curve, one has to understand the meaning of creating a new growth curve. A new growth curve will only exist when new market potential is existent, and one is able to respond to that on time with the right value proposition. Many ways of doing business, old technologies, suppliers, clients, and organizational forms should be questioned and being adapted or replaced by something new.

The Chinese central government showed by their 12th Five-Year Plan and other policies that they want to create a complete new growth curve for China. For Western readers, this might be bold and too ambitious, but for the Chinese, it is normal to make such big plans. In Europe and the USA, jumping the curves is more on a micro level of individual organizations; Nokia showed that they wanted to create consumer electronics enabling a better quality of life by connecting people as their slogan says. But Apple redefined markets and emulated Nokia. Jumping the curves is about being prepared for the future and being able to make offerings to future customers, what they ask for or want to have. This brings us to the subject of creating new business models.

5.4 New Business Models

A business model innovation is a combination of different single innovations leading to a complete new way of doing business within an organization or between organizations. New business models have the potential to create a new industry. The basic elements of a business model are presented in Fig. 5.3.

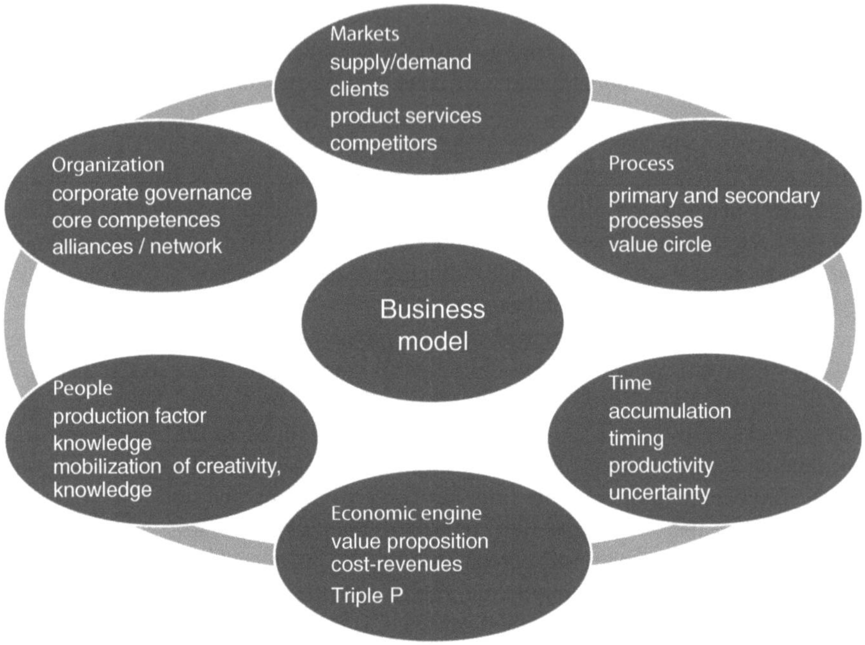

Fig. 5.3 Elements of a business model (van Someren 1991, Innovatie Emulatie en Tijd, Thesis Publishers & T.C.R. van Someren, 2005, Strategische Innovationen, Gabler Verlag)

Most of the business models can be described by the six basic elements mentioned in Fig. 5.3. But these dimensions are open for additional elements if needed. Per definition, innovation cannot be predicted with regard to timing, content, and impact. Therefore, it is impossible to give a complete definitive list of business model elements. They only help you in your search for renewing your business model in your specific situation and circumstances. For our purposes, short descriptions of the six basic elements are sufficient.

- Markets

  Of course the customer or client is the most important stakeholder. They determine demand criteria and future needs. But also the latent customer is relevant. Nobody asked for the invention of the personal computer, but when it was there, everybody wanted to have such a miracle screen. The next aspects are the products, services, or potential shift from products to services which makes the customer offerings concrete. Other stakeholders like suppliers, government, and public can influence the content and operational working of the business model. In case of sustainability, from a Western perspective, stakeholders get a larger stake in how business will develop.

- Process

  Every organization has its primary and secondary processes. A chemical company transforms substances into other chemical products to be used by B2B or

B2C customers. With regard to sustainability, the value chain (ranging from resource input, design, production, distribution, selling, and disposal) will be replaced by the value circle. The value circle, for example, closes material cycles and reduces emissions.

- Organization
  Organizational aspects can be manifold. The replacement of the profit criterion by Triple P within private companies requires a renewal of procedures for e.g. payment and bonuses and dealing with suppliers. But also alliances with external partners for co-creation of sustainable products and services can change the innovation function. Sometimes, new competences are needed to make the new business model work.

- People
  Traditionally, human beings are simple laborers carrying out a specific task. From a business model perspective, they can be regarded to be sources of innovation or have a different role in an organization. Human beings can also be part of the business model as the case of the supermarket shows. By means of the self-service innovation, customers were turned into costless laborers collecting their own stuff instead of being served by paid employees.

- Economic Engine
  The economic engine must prove that the new business model is economically viable and sustainable. This includes the value proposition and an insight in costs, revenues, and ecological and social effects. Hence, the economic engine includes the Triple P. Bottom line, the principle of value conversion holds for Western and Chinese situations. Any business activity needs in the end to lead to a value conversion of which a positive economic value is the minimum requirement. There are three determinants of value conversion: economics of scale, economics of scope, and economics of time (Triple E). For each case, situation, and period of time, these three dimensions can have different values and exist in different combinations. For an extended elaboration on Triple E, see van Someren (1991, 2005).

- Time
  Time is an underestimated variable in creating new business models. Time has different appearances but can have both a creating and destroying influence on the Triple P result. Of course time exists as clock time. Already this had an enormous impact on production organization as soon as time-motion studies in factories showed the influence on output and profitability. Time has a destructive effect when moves by competitors are not counterattacked quickly enough. Time as a productive factor appears when resources are invested in training. Sustainability requires a long-term view because the value circle ensures that inputs will come back as inputs with all their qualities.

The description or even categorization is left open for any additions for those who make their own interpretation. Every creative entrepreneur makes a unique interpretation and combination of these basic six elements which leads to a unique business model.

5.5 Context

The purpose of strategic innovation is not innovation itself but becoming unique and generating value. From a dynamic market perspective, being unique implies that imitation of your business model is very hard or takes a long time. Most technical innovations can be copied very easily or in a short period of time. A few exceptions exist like rocket or nuclear technology, but the great bulk can be imitated rather fast and with much less effort and resources than the original inventor. Hence, the sources of becoming unique do not stem from technology. The true sources of uniqueness and success are the nontechnical innovations and the context. By context we mean issues like external developments like megatrends, the level of uncertainty, cultural differences, and the way innovation is embedded in your organization.

Megatrends or a combination of trends and developments in fact imply future opportunities. But sometimes these trends can be created by yourself like Bill Gates and Ingvar Kamrad did. As soon as these (mega)trends are forceful or when core competences and economic engine fail, even the right to exist of an organization can be jeopardized. This also holds for whole industries, countries, or even economic regions. An example of the latter is the EU.

Uncertainty has an interesting impact on the way we deal with innovation (Knight 1964). An entrepreneur works in a context of high uncertainty due to a fast-changing world and because of the innovation itself. For an entrepreneur to cope with this uncertainty, small steps are almost a necessity in order to be able to reduce costs and risks, have a certain level of control, and to be flexible enough to deal with unexpected reactions and outcomes. An example is the influence of cloud computing. Cloud computing enables to organize against relatively low cost a pull market "on demand" based on the economic engine of "pay as you go or use." But not only technology inverts the market uncertainty. The introduction of the standardized McDonald's hamburger or Chiquita banana reduces uncertainty.

Moreover, cultural differences also have an impact on your view on innovation, uncertainty, and many more (van Someren 2005). For a German, innovation is hard-core technology, and its theoretical foundations, for an Italian, design is a very important aspect of innovation, and for a Chinese, innovation is practical technical applications. Also differences in risk perception have a cultural background. Americans are individual risk takers by striving for hero-like radical innovation and becoming rich. Japanese have a more collective attitude and prefer to reduce uncertainty by means of taking small steps after group consensus. These different views show that unique strategic innovations are not only dependent on the business model itself but much more on creating the right context.

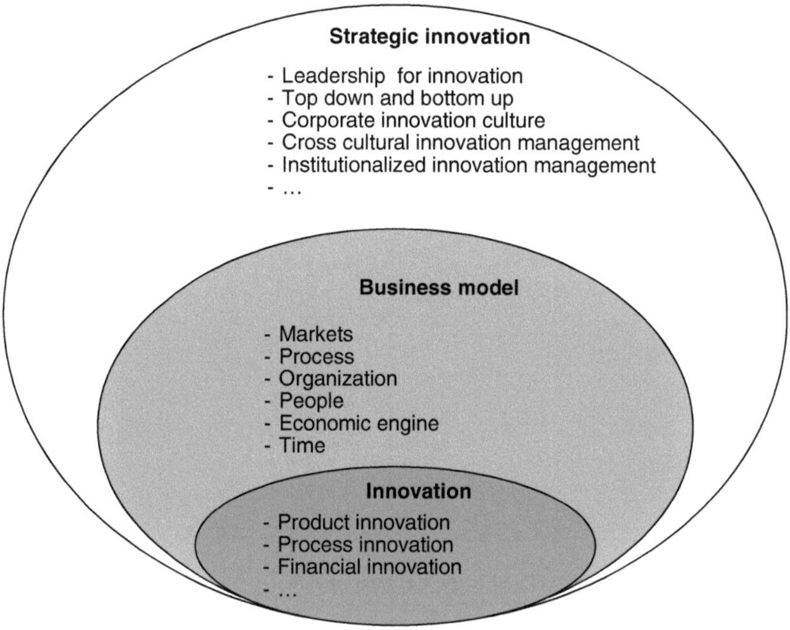

Fig. 5.4 The link between innovation, business model and strategic innovation. (Source: Based upon van Someren 2005, Fig. 3.6)

Therefore, jumping the curves in real life is not one big jump but a succession of small growth curves of small innovative experiments. The key to becoming unique is a combination of creating the right context and content such as a business model. Strategic innovation focuses on these topics.

5.6 Strategic Innovation

Strategic innovations include creating new corporations, new business models, new industries, new networks or clusters, or even new regions like metropolitan agro-food regions by integration of city and rural regions as one entity. Above all strategic innovation includes the creation of a context for innovation. You can hire a consultant to create your new business model but more important is to possess the ability to reinvent yourself on time. Strategic innovations refer to the ability to renew your business from within and to internalize and integrate the innovation into your organization or network of organizations. Strategic innovations not only include a new content in the form of a product innovation or a new business model but also comprise the ability, for example, to create an innovative culture, to demonstrate leadership, and to cope with intercultural differences. Only then jumping the curves is possible and staying ahead of your rivals.

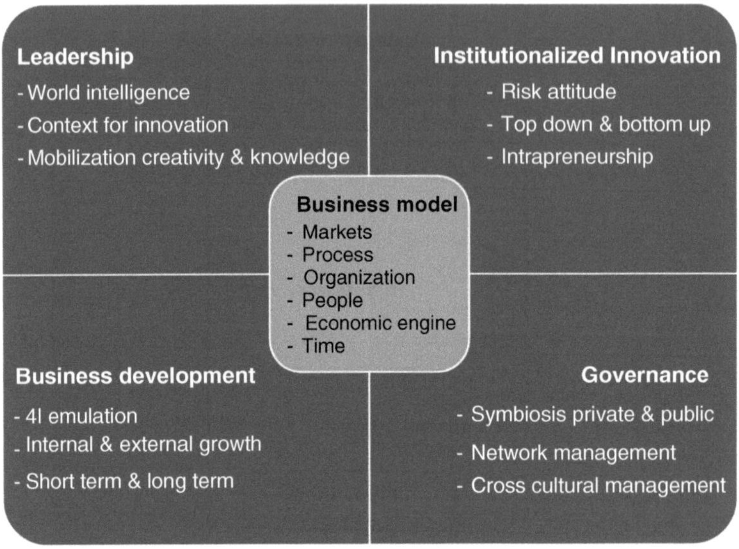

Fig. 5.5 Strategic innovation related to the business model

The basic relation between innovation, business models, and strategic innovation is presented in Fig. 5.4.

Examples of single innovations are product, process, financial, or ecological innovations. This category can be based on luck or ad hoc processes. A new business model is often a combination of several technical and organizational innovations. Strategic innovations include the context to be able to create single innovations and new business models on a regular basis. In general, strategic innovation includes technical and nontechnical innovations which can vary dependently on sector, geographical area, and time period. This renewal can also include or lead to new business models which are unique combinations of these elements of strategic innovation.

The ultimate purpose of strategic innovation is value conversion (van Someren 1991b, 2005). Depending on the scope of the definition, different approaches to value creation exist, such as value proposition (generating the idea) and value capturing (securing the value). Another variety is shared value creation, which indicates stakeholder involvement. However, these are academic exercises, and therefore for our purposes, it is sufficient to note that the innovative strategic activity of an entrepreneur leads to value conversion. Besides traditional financial or economic value, the notion of value conversion can comprise ecological and social aspects, forming together the Triple P (profit, people, planet). Value conversion can have a positive or negative value, and it can be quantitative (financial profit) and qualitative (social benefits, image benefits, or loss).

With respect to Green China, the difference between creating new business models and strategic innovation becomes extremely important. In Fig. 5.5, the top management issues of strategic innovation are placed in the context of business models.

We will shortly discuss these aspects of strategic innovation compared to the creation of new business models (based on van Someren 2005a, van Someren and van Someren-Wang 2011a, b, c).

5.6.1 Leadership

The most important task of top management is not controlling shareholder value but creating a context for strategic innovation. The nontechnical and less sexy measures are more important than stimulating a fancy R&D project. It is the most difficult task of any leader to create such an environment. Moreover, the international aspect of doing business also holds for generating and capturing the value of knowledge. Therefore, leaders need a world intelligence to identify, grasp, and exploit the worldwide opportunities. Business leaders need to prepare for the jump from ad hoc innovation projects to institutionalized innovation. The role of top leaders boils down to:

- Formulate tailor-made conditions and assumptions for institutionalized innovation which fits to the organization and their previous experiences
- Create a team which takes into account aspects as younger and elderly persons, man and women, diversity in cultural background, experiences inside and outside industry, and employees and other stakeholders
- Fostering intrapreneurship implying individuals and team to start businesses within the company or together with business partners
- Stimulating bottom-up processes for innovative initiatives
- Set up a reward system for successful innovation
- Allow failures but at the same time organize lessons learned

5.6.2 Institutional Innovation

Institutional innovation means the ability of an organization to generate innovation internally. This is more than a R&D facility. It implies an organization generating improvements, radical innovations, new possible business models, alliances with other stakeholders, and so on. It requires the mobilization of entrepreneurship, knowledge, and creativity within the organization or between related business partners.

The requirement is to allow bottom-up innovation combined with top-down governance. In practice, this is easier said than done, but it is one of the most rewarding and necessary organizational innovations to achieve institutional innovation. In a Western context, the core of strategic innovation is the renewal of existing business practices bottom-up, in which entrepreneurial activity is key (van Someren 1991b, 2005a, Wolcott and Lippitz 2010). This focus on bottom-up business models created under uncertainty is a crucial difference with the totally system-oriented, largely top-down transition management and innovation system renewals, which has clearly defined goals and sometimes projected technologies.

Creative entrepreneurs are the main drivers of strategic innovation, although the initial move can be made by new policies, for example, the zero emission laws in California state. Entrepreneurs start various initiatives, small and big ones, local and global ones, and now and later in time; the total effects of which taken together may in hindsight be called a transition or paradigm shift. This bottom-up process is not only limited to individual entrepreneurs but it is also possible within organizations. Supportive instruments are, for example, large-scale events or crowd sourcing. But then it requires a clear strategy, organization, and governance of the innovation process. Therefore, the ideal process is to align all parties and interests and to cooperate. Such a process can be true for a few parties and stakeholders and for a certain time period.

However, actual strategic renewal is a messy fight between parties in which cooperation and competition are intermingled, depending on their contribution to realize entrepreneurs' dreams. Within this perspective, a total transition of a certain dominant technical–social–ecological–institutional–political paradigm might be at the end the result of the consecutive strategic innovations. But it is certainly not the practical focus of entrepreneurs with regard to their own activities in their strategy and daily actions.

The risk perception within an organization determines the organization and scope of the institutionalization of innovation. Innovation is not about being open or closed but about the tradeoff between sharing risks and costs and grasping the benefits. This is dependent on the scale, scope, and time aspects of your business.

5.6.3 Business Development

Strategic innovation includes the creation of new firms or even industries and not only a new business model within existing firms. Therefore, external growth by means of alliances, takeovers, and mergers belongs to a business development. Moreover, external growth combined with a long-term perspective of achieving certain goals is a strong driver for success and emulating your rivals. This is something else than focusing on shareholder value which has long-term aspects, but these are too often overshadowed by short-term interests. Green China shows that dominating your business can be based on external growth and long-term goals of becoming Number One in the world. Green China uses the instrument of forced mergers between firms to achieve a desired level of concentration as carried out in the wind mill industry.

5.6.4 Governance

The governance of your organization or network also determines the process of value creation. The replacement of the profit criterion by Triple P is an example. But the different forms of cooperation between the public and private sector also illustrates alternative forms of governance. In Western countries, a common organizational

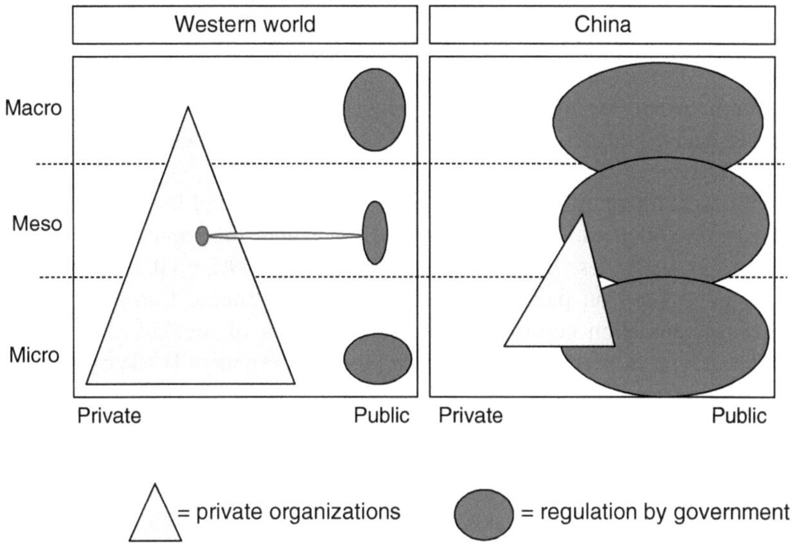

Fig. 5.6 Strategic innovation: Western world versus China

form of cooperation is the public–private partnership (PPP). However, these PPP constructions often have a loose form of cooperation compared to state-owned companies in China. The strong symbiosis between governmental organizations and business organizations in China shows successful outcomes. But this is not the only and best solution. Other solutions can work as long as they are carried out consistently. But in Western free-market situation, these forms of symbiosis prove to be very difficult and uncommon. In case of Airbus, the EU was able to set up a successful alliance between governments and business.

All these aspects demonstrate that fostering strategic innovation comprises more than the current hype of social innovation. Social innovation is an important aspect but only a relatively small part of the total picture.

5.7 Relevance of SIT for Green China

SIT offers a general approach for any organization in any industry in any country. In Fig. 5.6, the position of SIT in the Western world and China is presented.

It shows where strengths and weaknesses of both regions in the world are and in which direction possible future development can or should go. In Western countries within the market economy, the own initiatives of private organizations on a micro level have been the main driver of strategic innovation. On all levels, micro, meso, and macro, policies are available, but in most instances, they have no connection to each other in the sense of direct connection between government and ownership or direct support of private companies.

In the USA, the innovation policies focus on new technologies and the principle of market-based innovations. Principally, the approach in the USA is to reject or minimize governmental support or interference. But there are a lot of exceptions. For example, there are many linkages between the military complex and private industries. These budgets with direct or hidden support are beneficiary for airplane and IT industry.

The European Commission and its apparatus, central and local governments in member state countries, are too far away from where innovation is really taking place. This is Bruxelles. Bruxelles being the feared 100% civil servant city. The comitology focused on partial interests instead of strategic European goals and failing implementation contributes to a loose system of regulations and bureaucracy. Therefore, in Western situation, the problem is to identify the right approach to innovation and especially its implementation. What we need is a we try to express commercialization of Brussels instead of bureaucracy as an European commercial center. This is the true relevance of SIT for Green China.

For the Chinese side, on the macro level, the central government issues a Five-Year Plan which sets the broad guidelines and main priorities for the future. Although bureaucracy in China cannot be denied and also regions have their own interests despite central government, the link between national strategic aims, implementation, and private and public organization carrying out decisions is relatively tight. The central planning creates huge opportunities for industry, and if they have access to the right governmental channels, a fast growth is possible. This is a real win–win. Simultaneously, also individual sectors and firms will be chosen to take care of realization. On a meso level, the provinces are supposed to implement these plans but will also formulate their own strategic development plans. But also here, the symbiosis between government and private business leads to win–win situation.

5.8 Ideal Sustainable Development Company (ISDC)

A special application of the SIT is how to create a sustainable company or an Ideal Sustainable Development Company (ISDC) (van Someren 1995, 2005a). For most managers, it remains unclear what the sustainable corporation is about. To support them, the Ideal Sustainable Development Corporation (ISDC) can be described along seven archetypical characteristics. Achieving a sustainable world requires several strategic innovations. These seven characteristics are as follows.

5.8.1 Requirements of Stakeholders

Any organization is part of a society in which it operates. In order to survive and to be sustainable, the current and future requirements of stakeholders are relevant. Of course direct stakeholders like customers, employees, suppliers, and financiers determine future demand and supply requirements. But others like government, policy makers, NGOs, institutional organizations, international organizations, and the general public also have influence on the requirements.

However, there is another side as well. Only listening to stakeholders is one side of the medal. The other side is that organizations can take initiatives and launch innovative ideas without consulting the stakeholders. Henry Ford said: "If I'd asked customers what they wanted, they would have said 'a faster horse". No one will doubt this saying. But what do we think of similar behavior of Monsanto when they pursue GM food? In the eyes of many people the Monsanto strategy and its GM food is disputable. Should Monsanto listen to these stakeholders or continue with their GM food?

Also in China, R&D programs and experimental fields of GMO rice which need 50 % less water during the growing phase exist. Western companies are increasingly involving former "enemies" like Greenpeace and animal protection organizations. Due to resistance from European politicians, farmers, and public, beginning 2012, BASF announced to abandon their R&D and commercialization activities in Europe and relocate it to the USA (RSC 2012). This is not only based on philanthropic considerations but on co-innovation and sharing responsibility for the outcome.

5.8.2 Triple P Issues

Triple P will replace profit as the measure of business performance. Triple P is about social and ecological issues connected with any business activity, but creating economic value (profit) stays important. Therefore, sustainability is nothing else than creating new business opportunities and business cases, sometimes requiring a new business model (van Someren 1995, 1998a, b).[2] Identifying Triple P issues helps to make sustainability more practical. The logic first step is to identify all Triple P issues and effects of current business like CO_2 emissions, water sewage, waste, and so on. More important is, however, the future issues and the influence on future activities and the hidden opportunities. It will show the strengths and weaknesses from a Triple P point of view. Competitors and new entrants will try to replace natural resources within the existing business model or building up new industries. The latter is the most threatening proposition because emulation can become more difficult the longer the new propositions gains ground. Examples are electronic storage instead of paper storage.

Natural resources start with the design and the choices for certain materials being made. Traditional requirements for selection are, for example, manufacturing

[2] van Someren, 1995, "Strategies for Sustainable Profits," Milieumanagement van Kosten naar Baten, Wetenschappelijk Instituut voor Milieumanagement, Universiteit van Amsterdam, 7th April 1998a, Amsterdam, The Netherlands; "Strategies for Sustainable Profits", International BMW Seminar, Dietramszell, 17 September 1998b; "Strategies for Sustainable Profits", International Shell Seminar Managing HSE in the Business, 7th December 1998c, Noordwijkerhout, The Netherlands.

requirements, quality, and costs. Due to sustainability, others like recycling, lifetime, and resource depletion have to be added. In this respect, also availability and access to sources will be relevant. Energy refers to energy intensity, energy use of existing products and services, and of course renewables. Instruments like cradle to cradle offer opportunities for a sustainable product development.

5.8.3 Primary and Secondary Processes

Primary and secondary resources do not only have an ecological impact which can be measured by LCA techniques or footprints. From the perspective of innovation, primary and secondary processes can change dramatically as soon as alternative processes enter market. Producing and operating an electrical car leads to different primary and secondary processes which can alter the original equipment manufacturer (OEM) substantially. The agro-food sector shows how primary and secondary processes of different industries can be coupled, thereby reducing operating costs and ecological effects (van Someren and Nijhof 2010). For example, CO_2 from fertilizer production site can be used by greenhouse farmers. A step further is to create complete industrial parks where these principles are used. But the complexity and dependencies between the involved parties increase dramatically, and this is a yet unsolved problem how to deal with that.

5.8.4 New Value Proposition

Applying Triple P issues and new primary and secondary processes will lead to a new value proposition. New products, services, logistics, and ways of dealing with customers will enter the market.

Triple P value conversion changes the rules of the market game. Introducing green fruit or vegetables is on first sight a simple task. The contrary is true. The whole value chain has to be changed or even replaced by a new one. Fruit growers have to change their practice with regard to using fertilizers and crop protection. Logistical parties have to offer green transport services. Consumers have to be informed about the Triple P issues and the advantages and about its pricing. Very important in this respect is what is sustainable? Is it without using pesticides? Is it local production? What is local production? For a Dutchman, defining local would mean a maximum distance between production and consumption around 30–50 km. For a New Yorker, a distance of 1,000 km would still be regarded local.

Another aspect is fair pricing meaning a fair distribution of profit between farmers, logistical service partner, and retailer. In short, many practical issues have to be solved.

5.8.5 Time

Sustainability is based on a circular time concept. The notion of circular time is simple, its appliance has far reaching consuequences. The example of nuclear waste and CO_2 storage demonstrates this. Nuclear waste has a half-life characteristic which at the end will eliminate the nuclear radiation. It takes many years, depending on the substance, about 100,000 years, but the waste problems disappear. Some regard nuclear energy to be sustainable because it has no CO_2 emissions. As of 2000, the storage of CO_2 in old gas fields and other empty caves is seen by many as a solution for the global warming. It also represents a huge market volume because, worldwide, this technology can be applied. But CO_2 storage is in fact very unsustainable mainly because of the time effect. Why? The effect of CO_2 storage is extending the growth cycle of fossil energy sources (van Someren 2008). It hinders or at least postpones the development of renewables. Furthermore, CO_2 storage has to be maintained forever unless new yet unknown applications emerge. What are the costs of lifelong storage is a rhetoric question.

Other important time effects are timing and accumulation effects. When starting a new business, timing determines much more than any used technology or management instruments the bottom line results (van Someren 1991b). Too early and no profit can be made, and being too late leads to missed revenues which can never be retained. In addition of timing, the accumulation of experiences and knowledge also contributes to the time advantage. The timing effect of starting a business in renewable energy is shown by Chinese companies. Their early start gave them an advantage culminating in large firms. But the race is not over yet because new technology can start a new race.

Furthermore, the circular time concept has profound influence on the way value conversion is being realized. The circular time concept forces top management and policy makers to think long term. If they do not, large revenue streams will be missed and cost levels will rise. The circular time concept leads to the value circle replacing the linear value chain and the linear innovation model. The next section explains this phenomenon with far-reaching consequences.

5.8.6 Value Circle

When materials, energy, and primary and secondary processes are coupled with the circular time concept, the value chain will be replaced by the value circle (van Someren 2005). The value circle is much more than a notion; it has radical strategic consequences. When reusability, recyclability, or energetic value have to be maximized in order to attain as much as possible the ideal sustainable society, then design and choice of materials have to take into account these criteria. The cradle to cradle approach is an illustration of this. But there is more.

Location decisions of factories and recycle activities have to be measured against cost levels as well as ecological criteria. With the additional criteria of minimum ecological effects, the optimal geographical location decisions can have a different outcome. Production, distribution, recycling and reuse, and their new logistical flows based on Triple P criteria will change in choice of materials, location, and size of operation. Long-term decision taking is required because the nature of the products and their use determine their lifetime within the value cycle (van Someren 1995). When coming back from their use, the future decision makers confronted with the return stream of materials and have to integrate them in their own decisions.

This decision making is a complete inversion when compared to the linear value chain which can be described as a throwaway society. Therefore, in case of scarce resources, it is not only important to capture the mining activities and monopolize input factors and thereby securing manufacturing base. But the question who owns the returned materials and where are the recycling activities located becomes equivalently relevant. Capturing the mines and extraction of natural resources defends the manufacturing base, but who captures the used products and its recycling and reuse? Will it be near the manufacturing sites or will they be nearer to the location of consumption and point of return? Are they prey of the open market or can they be captured as well? And how is it done? These are strategic postponed time decisions characterizing the value circle, which were completely absent in the linear value chain.

5.8.7 Business Model

All the aspects taken together will in most cases require a new business model. For example, drinking water companies are based on large-scale and capital-intensive pipelines transporting water to households and businesses. With new technology like nanotechnology, a decentralized water production and consumption becomes a realistic option. Within the business model, the governance structure based on Triple P performance measurements has to be adapted. Unilever shows that in their attempt to create a sustainable business model, in the future, suppliers will have to fulfill Triple P criteria, and managers will be rewarded based on Triple P outcomes.

The ISDC demonstrates that a bunch of innovations are necessary to achieve a sustainable business model or even society. However, in relation to Green China, we know that we should focus on strategic innovation beyond the business model in order to become successful.

Take the case of solar PV. In 2000, the German government introduced the German Renewable Energy Act (in German, EEG) to stimulate renewable energy. In short, this law is based on a feed in tariff paid by the German consumers. His tax revenue is then spent to stimulate solar panel industry and make it affordable for consumers to buy PV solar panels. Soon new firms were founded with great prospects. However, Chinese overcapacity was exported to worldwide markets and in particular the German market. Based on low price competition as of 2010, many German manufacturers became difficulties, and some of them even went bankrupt.

The German perspective is that the Chinese intruders violate WTO rules due to dumping. According to German manufacturers, the low end price is below cost price. Some of the German manufacturers who have outsourced their manufacturing to China could not equal Chinese prices. The reality might be a little bit different. First, German manufacturers also get subsidies by the feed in tariff and often by local governments when setting up a new factory. It is estimated that in 2011, the solar PV contributes 5 % of total energy supply in Germany, but the system costs 100 billion euros of tax money. Of course Chinese manufacturers are indeed supported by their government. Second, the Chinese manufacturers had the strategy of high quality combined with economies of scale to keep costs low. Chinese government forced Chinese suppliers to merge in order to become a global low-cost player.

The structure of the supply side in Germany is totally different. Apart from some big players, there are many small-scale manufacturers with sometimes only 50 employees. Some of them tried to become a premium brand player that failed because the machines to make high-quality solar panels are also present in Chinese factories and delivered by same supplier. The result is that Chinese quality equals German quality. Another economies of scale effect is that many German manufacturers are listed on local stock exchanges like SDax or TecDax, whereas Chinese rivals get their capital from Nasdaq or even NYSE. Some German manufacturers choose a way out by outsourcing to China.

But producing in China has its risks and increases economies of scale of Chinese producer. Moreover, local German governments see their support vaporize. The economies of scale effect dropped prices. Moreover, in China, other appliances of PV cells were exploited, leading to economies of scope advantages. On top of that, because of high feed in subsidies, it became profitable to sell high volumes instead of investing the money in technological development. The strategy adopted by, for example, Solarworld to integrate the whole value chain under their roof did not work. In first instance, it was an advantage to control the whole chain, but the recession now makes it a heavy burden and low flexibility. With support of the Chinese government, decisions to close parts of the activities in the value chain can be taken from one day to the other.

Moreover, Chinese factories work 24/7 giving another economies of time advantage. The postponement (economies of time) of radical technology development, the low flexibility of some rivals, and the work around the clock gave the Chinese another time advantages. These are the true effects of Triple E (economies of scale, economies of scope and economies of time) leading to low-cost manufacturing. True global players get their capital from the US stock market, manufacture in China and sell world wide. For this reason, the Chinese are already able to produce high-quality products with low prices without subsidies. Taken together, these Triple E advantages together with cooperation with Chinese government are more relevant than the argument of cheap capital and dumping. Strategic innovation is more necessary than ever.

In the next chapter, we show several cases where the aspect of strategic innovation is the focal point in creating a new business universe.

References

BP Energy Outlook 2030. London, January 2012.

Knight, F. H. (1964). *Risk, uncertainty and profit, reprints of economic classics*. New York: Augustus M. Kelley.

RSC (2012). BASF pulls out of Europe over GM hostility. RSC advancing the chemical sciences, 18 Jan 2012.

Someren, T.C.R. van (1990). Organisational innovation. *Tinbergen Institute Research Bulletin, 2*(1), 65–73.

Someren, T.C. R. van (1991a). Emulation and organisational change. In M. Perlman & F. M. Scherer (Eds.), *Entrepreneurship, technological innovation, and economic growth*. Ann Arbor: University of Michigan Press.

Someren, T.C. R. van (1991b). Innovatie, Emulatie en Tijd. De rol van de organisatorische vernieuwingen in het economische proces. Tinbergen Institute Research Series, no. 9, June 1991b (diss.).

Someren, T.C. R. van (1995). Sustainable development and the firm: Organisational innovations and environmental strategy. In T. C. R. van Someren (Ed.), *Business strategy and the environment* (Vol. 4, No. 1, pp. 23–33). Jan–Mar 1995.

Someren, T.C. R. van (1998a). Strategies for sustainable profits. Milieumanagement van Kosten naar Baten, Wetenschappelijk Instituut voor Milieumanagement, Universiteit van Amsterdam, 7 April 1998, Amsterdam.

Someren, T.C. R. van (1998b). Strategies for sustainable profits. International BMW Seminar, Dietramszell, 17 Sept 1998.

Someren, T.C. R. van (2005). *Strategische Innovationen. So machen Sie Ihr Unternehmen einzigartig*. Wiesbaden: Verlag.

Someren-Wang, S., & Someren, T.C. R. van (2005). Succesvol zaken doen met Chinezen. *Wapen je met kennis van de Chinese zakencultuur. Fenedexpress, 289*, 15–17.

Someren, T.C. R. van (2008). In: Frank Staver, Trouw, Is CO_2-opslag struisvogelpolitiek? December 03, 2008.

Someren, T.C. R. van & van Someren-Wang, S. (2010). Revolutioneer het Transitiebeleid. *Milieu, 1*, 8–10.

Someren, T.C. R. van & Nijhof, A. (2010). *Triple P business development in the Dutch agro-food sector*. Assen: Van Gorcum Publishers.

Someren, T.C. R. van & van Someren-Wang, S. (2011a). Chinese strategie, in: Financieele Dagblad, 10 mei 2011, p. 7.

Someren, T.C. R. van & van Someren-Wang, S. (2011e). China speelt in coulissen grote rol in Durban. In: Trouw, 30 Nov 2011, p. 19.

Someren, T.C. R. van & van Someren-Wang, S. (2011f). Kans na harde landing. In: Financieele Dagblad, 2 Dec 2011, p. 7.

Wolcott, R.C. & Lippitz, M.J., 2010, Grow from Within. Mastering Corporate Entrepreneurship and Innovation, McGraw Hill, New York.

How to Generate Strategic Innovation for a Sustainable Future 6

Company of the Future in Practice: Green Car, Water, and Agro-food as Examples

Strategic innovation is at the core of the future Green China. But what does it mean in practice? A few cases show that in every industry or situation the appliance of the strategic innovation theory (SIT) leads to tailor-made answers. Only then unique business models can be created. It is also shown that the non-technological aspects of strategic innovation and creating new business opportunities are pivotal for success. Strategic innovation in practice learns that a new world can be made. Ten golden lessons are derived from the practical experience. Taken together they show that new ways of thinking, leadership, business development, client relationship, and so on are far away from business as usual. In addition, when applied in relation to Green China, the cases show that all the best resources have to be gathered to be successful. The implementation of strategic innovation represents the most difficult challenge for realizing Green China.

6.1 From Global Warming to Global Warning

We are warned louder and louder by sustainable issues and the consequences for our future welfare. There is no green revolution. But there is an evolution toward a green society. The first global warnings against the unsustainable characteristics of the industrial growth curve came in the 1960s. The report of Rome signalized scarcity of resources when extrapolating demographic development and industrial growth. Another highlight mentioning cracks in the existing industrial system and our current way of life came from the Brundtland Commission in the late 1980s. The Brundtland Commission came up with the notion of sustainable development as an alternative to the unlimited and linear growth models of most industrialized countries. Despite all these and other warnings, it took decades before some real changes toward a greening of society really took off. The warning was not always rightly interpreted. The first green wave reaching the top management board

T.C.R. van Someren and S. van Someren-Wang, *Green China*,
Management for Professionals, DOI 10.1007/978-3-642-28810-4_6,

meetings and governments started in the early 1990s. In Europe, environmental issues were top priority of some NGOs and pressure groups like Greenpeace, but much less these ideas were accepted by (top) managers. These ecological aspects were more or less regarded to be disturbing for running business. Some NGOs like Greenpeace took the road of hard action instead of paper work to direct attention toward overfishing of the world seas or misuse of natural resources. There were only a few governments in the West who took these reports and action seriously and tried to formulate new policies.

In Europe, some Nordic countries like Norway, Sweden, and the Netherlands soon followed by others like Germany came up with new environmental policies, laws, regulations, and voluntary agreements between private industry and governments. These regulations were focused on environmental compartments like water, soil, and air, or on industries like packaging or chemicals. However, the focus of governments, regulators, and pressure groups was solely one-dimensional on environmental issues. The external detrimental effects of the capitalist system were criticized. But there was no link between environmental issues and the market economy.

Of course, there is still discussion and uncertainty about the precise cause and effects of human action on climate. The physical effects of oceans, sun radiation on climate, sea level, and human health is largely unknown. From an ecological perspective, the dilemma is to wait for better information or to take action now because we cannot wait longer. From an economic point of view, there is in fact no dilemma. Investing in sustainability is often financially rewarding but it requires out-of-the-box thinking and accepting new realities. A simple example is: waste is not waste but represents renewed input and economic value. It was advocated for a long time to integrate environmental and social issues into the market process in order to make sustainability part of economic investment, cost–benefit, and other managerial decision taking (van Someren 1992a, b). Therefore, the measuring rod is not ecology sec but whether it makes economic sense and includes ecological and social improvement. This approach and interpretation has been chosen by the Chinese for creating and developing Green China.

Economic performance indicators needed to be complemented with social and ecological performance indicators on both operational and strategic levels of the company. But sustainability or sustainable development was not on the top management agenda of most private organizations. Most captains of industry had no clue what sustainability is really about or its relevance for their firm or industry. Only a few entrepreneurs like Mrs. Roddick and her famous The Body Shop were front-runners. But they had hardly any followers. There was no imitation hype like with Apple iPhone. In the 1990s, a shift toward managerial issues like environmental management systems, ISO 14000 and its European pendant EMAS, and reporting was initiated. The rationale was twofold. First, environmental problems do not only need technical solutions but nontechnical innovations as well. Second, prevention of environmental problems would be more cost and environmental effective than end of pipe solutions such as SO_2 or NOx emission cap levels. Early adopters of these management systems were the MNEs and chemical industry (van Someren 1992c, 1994, 1995b). The MNEs had more time and resources available to take care of new developments like environmental management. The chemical industry was leading the way because of some serious environmental hazardous accidents like with Union Carbide in Bhopal India killing hundreds of residents. But even then, sustainability was regarded to be part of middle management and often organized as a staff function. Core issues were low-hanging fruit issues like waste management, wastewater, or energy efficiency.

Seldom sustainability was interpreted as a strategic issue creating new markets and opportunities for new products and services. In the Western automotive industry, for example, it was almost impossible to go beyond the operational issues and question product portfolios, value chain resource efficiency, or new business models.

In China, where environment policies had to give way for economic growth for the last decades, a sustainable economy based on the principle of profit plus people and planet is being planned and tried out. The Chinese automotive industry is trying out green vehicles. What can the Western car makers do when there is not enough oil and road for every one? The repeated cry for a total solution for the urgent water problem in China did not lead to flourishing Western water companies. Agriculture and food sector in the West are earning money by feeding the Chinese. Can they see the future of Chinese metropolitan modern agriculture and food industry forced by their need for food security? Are we warned by the Chinese now that they can turn green to earn money? From the perspective of growth curves, the global warnings and the measures taken by government and industries did not lead to a new growth curve based on sustainability. On the contrary, managers tried to follow the easy way by integrating environmental issues into existing business. It was nothing else than improving the existent growth curve. In fact, sustainability was regarded to be an additional operational issue and not a strategic issue. The core idea to see sustainability as a strategic issue leading to new growth opportunities and outperforming competitors or even create new (sub)industries and accompanying Ideal Sustainable Development Companies (ISDC) did not end up in the top management meetings (van Someren 1992d, 1995a, b). For service industries like banking and insurance, it was a bridge too far (van Someren and van Dam 1997). But there were a few exceptions. In case of Shell and BMW, the two top managers globally responsible for sustainability accepted the idea of creating new market opportunities and new business models (van Someren 1998a, b, c). These managers agreed with the vision of sustainability being the foundation and start of a new growth curve requiring new business models. But their board of directors did not have the same insights and vision.

Unfortunately, sustainability as an issue was disappearing not only from the top management agenda but also from the radar of governments and policy makers. Another issue got far more interest of almost everybody. This issue was the Internet and dot.com hype. Boardrooms were filled with dot.com mania. Consultants were hired to work out the most hilarious business models without any economic rationale as long as Internet-based business was somehow center of new future activities. Bunches of top managers were blinded by Internet business and virtual profits which were in many cases only losses which in their mind would turn into gold bars. Once. At some time in future. Dotcommers as medieval alchemists. Many highly educated, trained, and experienced top managers lost sight of business risk and reality. Why? Short-term thinking, greed, and mass hysteria. Global warnings and global warming bothered almost no one. The first wave of sustainability ended. Furthermore, the Europeans were also occupied by their euro currency implementation. Some economic revolutions take time and sometimes an ignition from outsiders.

The green fire in the West was ignited by the global warming story of Al Gore with his film "An Inconvenient Truth." The dot.com heat fizzled out like a candle. Despite several scientific failures, Gore's movie showed better than any other data overloaded PowerPoint presentation the consequences of ignoring climate change. It put the issue of sustainability back on the global agenda of governments and policy makers. Now also some private companies were more seriously addressing this issue. But still sustainability had to compete with other business issues like the outsourcing, Internet, rise of Asia, and European integration. The UN climate panel with their Intergovernmental Panel on Climate Change (IPCC) reports followed by famous discussions about the truth of the data and objectivity of the researchers distributed attention to climate change and the supposed need for action. For the general public and most of the media, the second wave of sustainable development indeed started with Al Gore and IPCC. But the global challenge of greening the industry has taken different roads and approaches in the USA, Europe, and China. Western strategic green innovation does not equal the Chinese strategic green innovation.

Nevertheless, some front-runners started to realize that greening of industry was somehow necessary. As a result of these insights, a few companies started to implement the ideas and insights of sustainable development. But from the perspective of strategic innovation, we demonstrate that most initiatives were nothing else than improving the existing business model. A fundamental strategic renewal of the industry and creating a new growth curve were rarely tried. The cases show the attempts to make a breakthrough and make a jump to the second growth curve. Only since 2010, several firms try to fundamentally rethink their business model and start up the necessary strategic innovations to become a sustainable company. Most revolutions in business take time.

In China the green kickoff was ignited by the attempts of the central Chinese government to become a member of the WTO. These steps were taken in the early 1990s, and therefore, the Chinese timing is earlier than the timing of many Western countries. The Chinese government thought that when they would not pay attention to ecological and social issues, the accession to the WTO would be jeopardized. The Chinese government founded a specialized organization to take care of environmental issues in business and policies. The main goal of this new organization named Chinese Center of Environmental Management Systems (CCEMS) was the greening of Chinese industry.

Here our story of several cases in this chapter starts. Since 1991, your both authors were involved in the greening of industry and as of 1996 our paths crossed. From that moment on, we followed the same path, and our mutual and common experiences with regard to green strategic innovations are presented here. From our own experience in the early 1990s till now, we know that when bringing up the strategic challenges of sustainability at top management, no serious reaction came. Even the rise of China was more regarded to be an extension of the existing business instead of a new growth curve with different rules and a different context. Only a few clients really bothered about developing their strategy around these themes. But with the strategic innovation theory and the special case of the Ideal Sustainable Development Company (ISDC), an instrument is available to find your path. We present several cases of introducing sustainability as a business case to show success, failure, opposition, disbelief, and above all the long time needed to success.

Of course, it is impossible to tackle all our assignments. Instead, we discuss some typical cases which illustrate four aspects of Green China. First, strategic innovations as described in Chap. 5 are necessary to create and implement sustainability into business practice. Second, strategic innovations can be applied and are relevant for both private as well as public organizations. Third, Western and Chinese approaches to sustainability differ in aim, scale, scope, and speed. Fourth, because of differences in conceptual thinking and approaches of solving and tackling the problems, a lot of business opportunities for greening China are being missed by Western firms.

The cases have deliberately selected from industries which will be crucial for the future of China. These industries are automotive, utility (water), and agro-food. The cases show the relevance of trends, the interpretation of opportunities, the existing and needed competences, the bottlenecks for radical and fast change, the business cultures and difficulties to create, and the implementation of strategic innovations. Seeing the right trends or combination of trends is not sufficient; one needs to create, for example, the right organization for renewal, foster innovation by a stimulating culture, and have the creativity and insight to be able to build new business models. Success is dependent on a large number of right and aligned decisions.

The relevance for Green China is that these kinds of radical strategic innovations will be crucial for future prosperity. The imitation-improvement scheme which has led to the current success of China will be insufficient. Generating strategic innovations is the big challenge. The cases show the different attitude and out of the box thinking needed to jump to new growth curves which go beyond technology. Future Green China can learn from these cases.

6.2 Green Cars

The automobile industry is booming again after the dip caused by credit crisis. They were saved by the Chinese. The German economy could grow the last few years because the Chinese were buying from them. However, if we look at the mega trends, we will see that the German will not laugh over a few years when the Chinese roads are jammed everywhere, and the oil resources are limited. In fact, the signs of a changing car demand are already visible. Chinese government aspires Chinese companies to be sector leaders in green cars. As a consequence, in the future, Chinese government will stimulate Chinese brands by all kind of policies like procurement criteria and market presence. Sooner or later, the traditional car industry will be replaced by a green one. Creating a green car industry confronts Chinese management with the creation of strategic innovations as Western industry has experienced in the past.

In this case, we demonstrate a few highlights of the long path of mainly the European attempts in the automotive industry toward becoming a sustainable industry. Thereby, we focus on our assignment at BMW.

6.2.1 Merging Instead of Greening

In the early 1990s, sustainability was not on the corporate radar of most car manufacturers. The reason is that most European firms had a restless time with Toyota's innovation of Just-In-Time. It took the European automotive industry about 20 years to identify, understand, imitate, and adapt the organizational innovations of Just-In-Time to their own ways of organizing and managing. Some industries like Mercedes Benz and BMW were in the mid-90s also involved in creating larger companies. Mercedes Benz had the vision of becoming a World Company ("Welt AG") by merging with Chrysler and an alliance with Mitsubishi. BMW also desperately tried to increase scale above the perceived minimum level of one million cars output a year. Other European companies like VW, Renault, and Fiat were struggling with the aftermath of Just-In-Time and getting their cost efficiency and quality at acceptable levels. BMW wanted to merge with British Rover. Meanwhile, Toyota introduced their first commercial hybrid cars. Both Mercedes Benz and BMW had first to digest their disastrous mergers before paying attention to a not-invented-here hybrid propulsion system. The merger with Rover nearly destroyed the company and left them over as a takeover candidate. For many years, the best managers of BMW were involved in time-consuming pre-takeover issues, post-merger integration, and disentangling the two companies. Simultaneously, BMW spread out in the world setting up production facilities in North America and South Africa. This internationalization was a major step. Before the merger, within BMW, German was the leading language, and now also middle management had to speak English. During international meetings with senior and middle management, one of our respected tasks besides being strategic sustainability consultant was to translate discussions from English into German and the other way round. In this time period of the early 1990s, BMW started to intensify their attention to sustainability issues. Both Mercedes Benz and BMW developed their own versions of hydrogen cars. Despite several pilot tests like BMW's hydrogen buses at Munich airport, these research projects never were really commercialized.

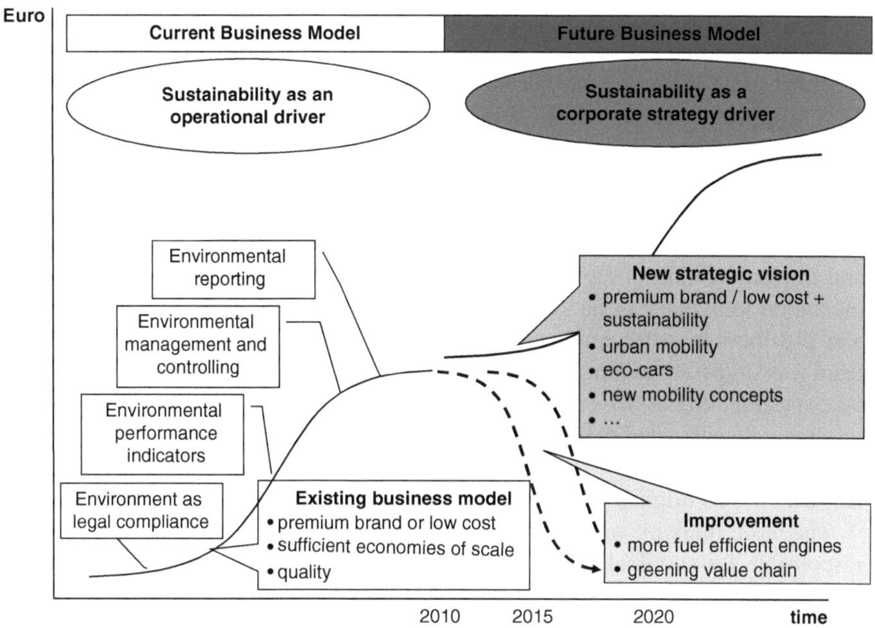

Fig. 6.1 Green strategic innovation in European automotive industry. (Source: T.C.R. van Someren, 1998b)

6.2.2 Sustainability as a Strategic Driver

Sustainability as a top management issue almost started from scratch, and successively several issues and projects were started and implemented. At that time, a senior top manager for sustainability vigorously initiated and stimulated ecological and social issues. To this top manager, it was clear that sustainability would be in the end a strategic issue. But first, existing operations like production, waste management, procurement, and distribution had to be taken care of. Fig. 4.1 shows a few examples.

In Fig. 6.1, only a few major organizational innovations are drawn, but along the curve of the existing business model, also a bunch of technical innovations were implemented. Examples of these technical innovations are recycling and technical waste and emission reduction solutions on several locations in the primary process. As can be seen, the majority of the issues are operational issues, or in our words, or an improvement of the existing business model. Of course, this is a logical course of developments, but one should keep in mind that it is in the same time period when Toyota started to develop its now famous hybrid technology. BMW had at that time their H2-car project, but it was more a greenwashing project than a serious attempt to change the rules of the business. Traditional managers with traditional consultants dominated the market, resulting in redesigning premium brand strategy and rising minimum output scale. Also in BMW, sustainability issues were organized over many different although not coordinated staff departments and line responsibility. Therefore, the traditional sometimes revised corporate strategy remained based upon the premium brand concept. Of course, Mercedes Benz was regarded to be the archrival, but now some new stars like Audi and in particular Toyota appeared. Around 2000, some senior managers believed that not Mercedes Benz would be the main rival but Toyota. Sticking to the premium brand strategy implied that sustainability issues should be incorporated into the existing business model. As Fig. 6.1 shows, all kind of operational issues like auditing, training, waste, emissions, green procurement criteria, ISO/EMAS certification programs, and implementation of environmental management systems were tackled. Implementation took place in factories and other operational units like R&D, planning, and distribution. After implementing all these operational successes, the BMW team was suggested to jump the curves and to rethink current business model and to take sustainability as driver of a radical new strategic vision. In a top management session, we confronted managers with the upcoming opportunities to rethink their business model based on sustainability as the main strategic driver. Figure 6.1 is identical to this original sheet of this proposal.

The basic thinking was supported by the top manager and his worldwide team responsible for sustainability. It was presented to board members responsible for sustainability. The outcome of the discussion was mixed. A few high-ranked managers believed in a radical change, but others rejected the idea. Even the former CEO, Pischetsrieder, commented the hybrid engine of Toyota as a bad solution because two engines in one car could not be efficient. Later on, in 2004 as top manager of VW, he argued that hybrid technology from an ecological perspective is useless and the wrong path. The reasons are that production of hybrid cars is ecologically very inefficient, batteries are chemical waste, and fuel efficiency is only better under very specific circumstances.[1]

In that time period, all the competitors which we spoke like Daimler, VW, GM, and Ford had the same line of strategic thinking. Moreover, the new market segment sports utility vehicles (SUVs) with very high-profit margins started to explode. The big profitable sales volumes of their SUVs and the race for more horsepower proved the correctness of their strategy, and sustainability was at the most an operational issue to be tackled by environmental departments. Short-term demand for profitable big cars like in the USA blinds management for a trend toward eco-cars. Sustainability was for a long time regarded not to be sexy and not very stimulating for making a career in the automotive sector.

Furthermore, the marketing success of their Efficient Dynamics program is a well-orchestrated improvement of the existing business model. These staff departments had no strategic power and were not listened to as soon as corporate strategy was at stake. Even worse, in the beginning of the new millennium, BMW was leading the Dow Jones Sustainability Index (DJSI) as best in class for years. The paradox situation emerged that the Board thought that they were performing

[1] Wirtschaftswoche 21.01.2004, Toyota setzt auf die neue Technik: Hybrid-Antrieb ist für Pischetsrieder Irrsinn.

well even in the area of sustainability, whereas from a strategic innovation perspective, a more bold strategy was necessary. As a result, the corporate strategy remained to be based upon the premium brand strategy to which environmental issues were added. About 10 years later, the lost terrain had to be catched up together with rival Mercedes Benz.

Image and window dressing dominated Western strategies. China was integrated by selling and exporting some products and outsourcing the manufacturing function. Any commercial organization is seeking new opportunities for cost reduction, new revenues, and profits. But at the same time, organizations are captured in their own historical and current success.

BMW entered the Chinese market rather late. Since the 1990s, it seriously started to build up its Chinese organization. Long before BMW, VW was the first European car manufacturer with substantial business in China. Based on its smart strategy to develop the market segment of official government cars, a large volume could be achieved. In the beginning, this was not an option for BMW with their premium brands.

About 10 years later, the top management of BMW showed publicly their new strategy in which the premium brand strategy was still the core but also sustainability was one of the pillars. Interesting to note that meanwhile, not Mercedes Benz was regarded to be the main rival but Toyota, and also Audi managed to become a premium brand manufacturer. BMW needed a strategic alliance with Daimler for sharing the huge development costs to develop hybrid technology. The new corporate vision of BMW as of 2008 is to be the leading provider of premium products and premium services for individual mobility. At the end of November 2010, BMW announced a cooperation with Toyota in which Toyota gets access to small diesel engine technology and BMW to hybrid technology.

The successful efficient dynamics strategy is an excellent illustration of improving the existing business model instead of renewing it. Hybrid technology was rejected for a very long time. But since market demand and governmental support for hybrid technology makes electrification of the engine unavoidable, in 2010 BMW teamed up with Mercedes Benz to develop this technology. With the introduction of the i-series, small vehicles for urban mobility, BMW tries to exploit the trend of sustainability and urbanization.

There is definitely a potential for the Chinese market. But it depends on how BMW develops this strategic innovation. In all their negotiations with Chinese alliance partners, in each round, and for every enhancement of their factories, they had to hand over technology. One of their rivals, BYD with a very look-alike BMW logo, is the main battery manufacturer in China. Is BYD trying to be the dominant electrical producer of China and cooperates with Daimler? This case shows that timing and accumulation of knowledge in new areas is crucial. For China, Electrical Vehicles (EV) offers a big opportunity to emulate Ford, GM, Toyota, Mercedes, BMW, Audi, Renault, Nissan, Honda, and all others. But they will leapfrog in another way as we will see below.

The right timing has a great influence on accumulated profit and losses or net present value over time. In case of BMW, the premium brand strategy is hollowed out by the rise of competitors like Lexus and Audi. More important are the new entrants like Tesla, Fisker, and BYD who directly move into the electrical car market. Especially, BYD has to be watched because they were originally the world's largest battery producer who moved in only recently into automotive sector.

It became also clear that slow adoption makes you a follower and not a business leader. Sometimes it is better to have a fast follower strategy, and thereby avoiding first mover investments and costs. But if you pretend to be a premium leader with focus on engine technology, then one has to defend its position and to be first mover. But why do companies hesitate or move slowly? There are several reasons. Management of traditional car companies has to create short-term shareholder value which forces them to defend existing profitable business model based on current technology.

Investing in new technologies like electrical cars require new R&D, new HR competencies, new production facilities, new supplier and procurement organization, and so on. Huge investments are needed to develop the radical new propulsion technology which have to be earned in traditional business. Existing OEMs cannot survive by selling electrical cars; there has to be a tipping point for a switch to new technology. The transformation from traditional manufacturing toward a sustainable manufacturing has again been emulated by an improvement in new products. The hybrid car of Toyota was first received with a lot of sepsis by dominating leaders in the automotive industry. Others like BYD, Better Place, Tesla, and Fisker innovate the propulsion technology or new car concepts by introducing electrical cars. Others like GM with their Volt soon followed. Most incumbents have not made real bold strategic moves from being a car manufacturer of something on four wheels toward, for example, becoming a supplier of new mobility services. It shows that revolutionary changes within incumbents are seldom initiated.

The Triple P bottom-line results of all the efforts of the first waves toward becoming a sustainable company is shown in Fig. 6.2.

The Triple P shows the improvement characteristic of the automotive industry. In case of BMW, the premium brand strategy is maintained as core of their business. Premium brand now includes sustainability. Driven by international agreements, national governments, and European policies, the focus is on emission levels particularly CO_2 and health and safety regulations. But traffic congestion and urban mobility concepts especially in mega cities in emerging countries are still unsolved.

6.2.3 From Mega Trends to Sustainable Car Industry

At the moment in the automotive industry, it is still undecided which new technology will dominate. The three main alternatives are biofuel, electrical, and fuel cell technology. All problems boil down to solving technical problems in energy source, safety, cost level of sourcing and production, and using the vehicle, consumer acceptance, and required infrastructure. For new market entrants, with regard to their technology choice, this is different because they have high R&D costs but

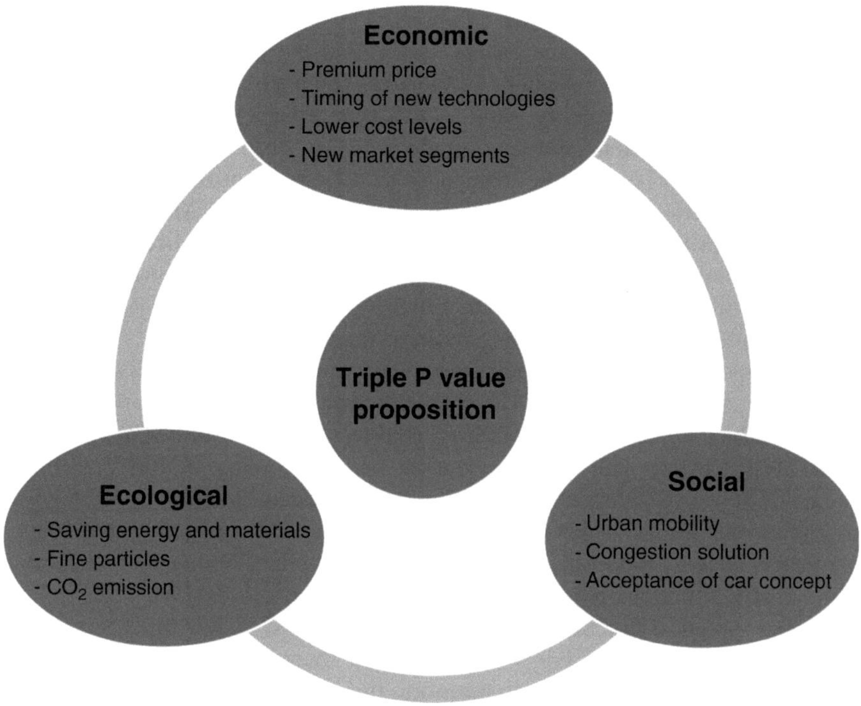

Fig. 6.2 Triple P at BMW

lower indirect costs. If successful, their need for investment will flow in technology improvement, scaling up, production, distribution, and marketing. But they accumulate experience, knowledge, image, and brand name. Lacking infrastructure for electrical cars does not contribute to creating short-term shareholder value. The accumulation of knowledge, experience, patents, and (brand)image can be decisive as soon as market developments accelerate. This can be seen at BMW who need alliances with Mercedes and Toyota to keep up with rivals in hybrid technology.

The Triple P in Fig. 6.2 demonstrates the rather evolutionary changes toward a sustainable company. The slow change has two sides of a coin. The first is that huge investments in new technology in almost every activity and function of the company is necessary. Old competences in R&D and manufacturing for producing a gasoline car have to be replaced by knowledge in electrical engineering. Equally important is the payback time and net present value of these investments. The long lead times for development and commercialization of a new car require years.

Additionally, the return on investment is also dependent on investments in infrastructure for the replacement of gas stations by electrical load points. Here the automotive industry is dependent on others like government. The second side of the coin is that existing enormous investments in technology, competences, and image will be defended by both the incumbent automotive manufacturers and the oil companies. Therefore, it depends on the speed of growth and development of newcomers, the acceptance of their solutions by customers, the strategic innovation capacity of existing firms, the energy sector, and the government how fast the European automotive sector will be sustainable. The main challenges for the future are reinventing urban mobility, services that span a vehicle's lifetime, integrated safety, clean production, and new products.[2] The sustainability strategy is to become the most sustainable company in the automotive sector. Still the sustainability strategy is separated from corporate strategy. Meanwhile, BMW has launched its BMW Efficient Dynamics program and has launched a new vehicle program to develop a megacity car powered by electricity called BMW i3. Within the context of strategic innovation, the Efficient Dynamics program is nothing else than an improvement of the existing business model.

Furthermore, BMW has founded a new brand with an alternative driving concept specific for the Chinese market. Together with their alliance partner Brilliance, this brand will be pure Chinese but will have no direct relation with BMW. According to BMW's CFO, it will therefore not be build according to BMW technical standards otherwise dilution with BMW brand will occur. In this way, BMW tries to circumvent a total technology transfer based on Chinese policy that foreign carmakers have to cooperate with Chinese firms and develop together cars. As of 2013, BMW will globally introduce its own new electrical "BMW i".[3]

In our proposal for strategic innovations options, a few truly innovative alternatives far out of the box were formulated. A bold option suggested in the proposal of strategic innovation to transform from a manufacturer toward a mobility provider was acknowledged as interesting but also rejected. The idea was based on the core idea that not the possession of the car is core of the business but the use of it. On average, privately owned cars are parked and stand still for far over 90 % of their time. From an ecological point of view, it is much more resource efficient to share this car with others. In fact, this strategic innovation would transform the company into a mobility provider instead of staying a premium brand OEM. Of course, this is a bombshell under the existing business model of maximization of manufacturing and selling cars. But if you do not do it, somebody else will try it.

Table 6.1 highlights some key challenges for the Chinese automotive sector. The biggest challenge is to transform from a copycat to an innovative sector and adapting to foreign sellers markets. Green China is beyond car manufacturing and green tech. Due to increasing urbanization, the rising income levels, and the growth of the middle class, the demand for traditional cars exceed the capacity level of traditional car concepts including e-cars in megacities. Within the current system, CO_2 rights will be the core of new mobility solutions. Mobility service concepts beyond the car are necessary. The challenge of Green China is to create the new sustainable growth curve. Furthermore, transparency and communicating with NGOs about corporate policies and performance are the soft skills side of

[2] Based on: Annual Report 2008, discussions with BMW representatives during 2009 and BMW Group website June 2011.

[3] Spiegel.de, 11.09.11 BMW Gründet neue Automarke für China.

Table 6.1 Chinese management agenda for the car industry

Megatrend	Issues	Examples of challenges
Closing material cycle, diminishing resources	Value circle	Creating recycle industries for car parts and material
Green mobility, rising income levels, larger middle class	Strategic innovation	Mobility services beyond car manufacturing, CO_2 rights based competition, focus on nontechnical innovations, new leadership styles, (inter)national R&D and innovation centers, Triple P governance, international alliances, standards of new technology, integration of Western cultures within organization, bottom-up innovation process
Urbanization	Business model	Development of design, new urban vehicles, alternative technology development, efficient manufacturing with high EHQS (environment, health, quality and safety) standards, differentiation between Chinese and customers from other countries
Climate change, CO_2 as constraint	Sustainability	Fuel efficiency, smart traffic systems, mobility services, transparency in performance
Globalization	Operational	Sales organization, service centers, branding, marketing, HR, procurement, customer relationships

sustainability. Of course, the improvement of existing Chinese manufacturers in traditional management areas such as branding, service, and global competition is still relevant for the coming years.

6.2.4 Lessons Learned

Lesson 1: New trends, early warnings, and weak signals like sustainability have been underestimated. The big jump to electrical cars was indeed too far due to insufficient technology and failing infrastructure. But the intermediate step of hybrid technology was underestimated in its potential despite commercial success of one of the main competitors. Apparently, these very strong market signals were not able to act as a wake-up call.

Lesson 2: The wake-up call was missed not because of failing information but due to technological tunnel vision and sticking to conservative premium brand strategies. The corporate strategy of greening existing growth cycle is nothing else than a typical technology improvement strategy.

Lesson 3: The right timing proves to be the most difficult strategic decision depending on insights in technology, market demand, infrastructure, and renewal of industries. Incumbent companies with large-scale operations need to be careful with jumping to new growth curves because only large markets can secure existing size of activities. Because of this economies of scale argument, SMEs and innovative new entrants can take the lead due to earlier timing and accumulation effects. The former non-premium OEM Toyota becomes a premium hybrid OEM and beats current premium brands such as BMW, Mercedes, and Audi.

Lesson 4: Within the established players, the investments in existing technologies exceed the investments in new technologies (exploitation versus exploration beyond green washing). This investment behavior opens the door for small flexible newcomers focusing on new technologies.

Lesson 5: The accumulated investments and experiences of core competences in combustion engines hinder the full-scale transformation to a complete new technology and business model.

Lesson 6: The institutionalized innovation system has been focused on short-term improvements and largely ignoring long-term commercialization of radical innovation.

Lesson 7: In the end, the economic dimension of the Triple P is decisive about timing, growth, and development of sustainable industries. Therefore, ultimately the Triple E of both old and new industries as well as incumbents and newcomers decide about winning strategies and not the Triple P.

Lesson 8: New technologies like hybrid, electrical, and fuel cell cars still build on the traditional concept of mobility on four wheels. Nontechnical innovations like creating new mobility concepts for megacities or combining public and automobility go beyond car manufacturing and alternative propulsion concepts.

6.3 Water Cycle Company

Water is a crucial issue for China. In the eyes of the Chinese, the Germans are the benchmark in the automotive sector; for water management, the Dutch are at least very highly regarded. Food and water in China have tremendous problems with health, safety, and quality issues. The water cycle company is a Dutch case showing the appliance of strategic innovation to cope with the challenges of climate change and providing drinking water at affordable price levels and high quality.[4] Water management is only for a small part determined by technology. Equally important is the institutional structure of the whole sector and the aims and capabilities to reinvent existing organizations. Also here, Green China is more than acquiring and showing off the newest technology.

6.3.1 The Fragmented Dutch Water Sector

On 1 January 2006, the first water cycle company was established in the Netherlands: Waternet. To understand the reasons and arguments for starting Waternet, responsible for all water services in and around Amsterdam, it is necessary to know the organization of the public water sector in the Netherlands and to know the challenges that face the Dutch water sector. The Dutch public water sector has a long tradition. Institutionally, there are three main players.

[4] This section is largely based on an earlier publication of van der Hoek et al. 2011, Integration and Innovation of the Urban Water Cycle: The Waternet Experience, in: Journal of Environmental Science and Engineering, 5 (2011a), pp. 533–544.

The water boards are responsible for safety against flooding, storm water management, quantity and quality management of surface water, and treatment of urban wastewater. At present, there are 26 water boards in the Netherlands. Drinking water supply is the responsibility of ten publicly owned limited liability companies. Shareholders are municipalities and provinces. In the last 30 years, a consolidation took place for quality reasons and efficiency reasons, resulting in 10 drinking water supply companies now. Sewerage is the responsibility of the municipalities. At the moment, there are 431 municipalities in the Netherlands, all having the responsibility for sewerage. The turnover of the drinking water companies is euro 1,413 million, the turnover of the water boards is euro 2,264 million, and the turnover of the municipalities (sewerage) is euro 1,006 million (Vewin 2008).

This institutional setting in the Netherlands has had a tremendous positive effect on public health, environmental quality, reliability, and safety. At the same time, changes in society and technology take place, resulting in challenges for the future. Hence, the question is whether this tripartite approach is suitable to cope with the challenges we are facing now.

6.3.2 New Challenges

Society is confronted with new trends and challenges, which may require new water management concepts. A European research project, TECHNEAU, has identified 10 trends for the Netherlands (Segrave 2006), ranging from technological trends to demographic trends, the climate change and the water cycle; business culture, the request for maximum efficiency; technological innovations, nano- and biotechnology; globalization and emerging markets (India, China); sensors, tele-detection and ICT; individualization and sociocultural integration; the energy crisis and mobility/transport; safety, trust, and terrorism; society's interest in the environment, sustainability, and healthy living; aging population. Many of these trends affect the water and wastewater services and are challenges for the water sector. Especially water-related challenges like climate change, a new European water regulation such as the Water Framework Directive and the need for efficiency in water management may ask for new water management approaches.

6.3.2.1 Climate Change

Climate change will have a very pronounced effect in the water sector (Moll 2007). River discharges will show more extreme values with lower values in winter and higher values in summer, resulting in larger quality and temperature variations and even periodic shortages of water with sufficient quality for drinking water treatment (De Bruijn and Van Mazijk 2003; Zwolsman et al. 2007). Adaptation and development of new concepts, including water storage, are required in the future. It is estimated that 50 % of climate adaptation costs will come from adaptation measures in the water cycle. Furthermore, mitigation by reducing or eliminating greenhouse gas emissions is necessary. At a European level, the target is to reduce the emission with 20 % in 2020 compared to 1990. The Dutch government has the ambition to realize a reduction of 30 % (Van Dorland et al. 2009), while the city of Amsterdam aims at a reduction of 40 % in 2025. Although the water sector has only a very small contribution compared to other sectors like energy production or mobility, reduction of greenhouse gas emissions in the water sector is important as all sectors are working on this topic. In the Netherlands, the contribution of the water sector to the total global-warming potential is 0.8 %, while for households, the yearly greenhouse gas emission caused by water-related activities is 3.3 % as compared to the emission caused by energy use (Frijns et al. 2008).

6.3.2.2 The European Water Framework Directive

The European Water Framework Directive which came into force in 2000 aims at an improvement of the water quality, both chemically and ecologically (European Union 2000). In order to reach this objective in 2027, specific measures and substantial investments are planned in the coming years.

Although the water quality has improved since the introduction of the Dutch Directive on Surface Water Pollution in 1974, still the water sector has to deal with many new emerging substances which are a threat for the water quality, like pharmaceuticals, endocrine disrupting compounds, and new plant protection products (Van den Berg 2009; Sacher and Stoks 2003; Mons et al. 2000). What kind of measures is necessary to solve the problems of emerging substances? Do we need more control at a source to avoid pollution of the environment with these substances as the drinking water operators argue, or do we have to remove these compounds at dedicated locations, that is, in the wastewater treatment plants or in the drinking water treatment plants? Where do we have to take actions in the water cycle to provide sustainable solutions?

6.3.2.3 Efficiency

From several studies, it can be concluded that the costs for drinking water, wastewater treatment, and water management in the Netherlands are acceptable (Geudens and Van Beek 2007; Unie van Waterschappen 2009) and relatively low in an international perspective (North European Benchmarking Co-operation 2008). Especially in the drinking water sector, important efficiency improvements have been realized (23 %) since the introduction of benchmarking (Geudens and Van Beek 2007). At the same time, policy makers argue for more efficiency. The position of the Dutch public water sector as a monopolist and thus the absence of market-based competition may feed this quest for efficiency. In a recent Dutch policy report, "BestuursakkoordWaterketen" (Policy Agreement Water Cycle), it is argued that an efficiency improvement of about 10–20 % is possible within the next 10 years (Ministry VROM 2007). The report also gives the direction in which this efficiency improvement can be realized: more cooperation in the water sector and a more coherent policy in the water sector.

6.3.3 A Response Strategy: The Water Cycle Concept

All these developments and challenges have one thing in common: they are characterized by great uncertainty, the problems underlying the trends are complex, and many stakes come into play. The future is complex and dynamic. It involves general trends and unexpected events in both the water and the social system, necessitating new transdisciplinary approaches (Haasnoot et al. 2009). Society aims at sustainable water management, which means that it is effective, robust, and flexible (Offermans et al. 2009). Drinking water supply, wastewater treatment, and water management are all affected and show interaction and interdependence with respect to the discussed developments. The institutional setting with three different organizations, each having only a task and a responsibility in a part of the total "water cycle," involves the risk of suboptimization and suboptimal solutions. In addition, the point has been reached where the separate elements of the water sector have been optimized and have reached their maturity. For example, in the wastewater treatment, the carbon removal efficiency is 92 %, the nitrogen removal efficiency is 80 %, and the phosphorus removal efficiency is 82 %. More than 99 % of the treated wastewater effluent discharges comply with the discharge permits. More than 99 % of the households are connected with the sewerage, and 100 % of the households are connected with the public drinking water supply. In the drinking water treatment production processes, water losses are less than 4 % (Vewin 2008). According to the yearly evaluation of the drinking water quality by the health inspectorate, the quality is characterized as good without any negative health aspects due to noncompliances with the regulations (Ministry VROM 2009a). Economically, the drinking water sector has realized an efficiency improvement of 23 % in the period 1997–2006 (Geudens and Van Beek 2007), while in 2007 and 2008, an additional improvement has been reached, resulting in a total efficiency improvement of 26.3 % over the last 12 years (1997–2008) (Vewin 2010).

To find an answer for all the new challenges, it is necessary to look for the interaction of the separate elements and use the interfaces. A more holistic and integrated approach is necessary for sustainable solutions in the future.

The water cycle concept is an approach that offers this holistic view. Drinking water treatment and distribution, wastewater collection and treatment, and water system management and control are considered as an integrated system. Because the water cycle integrates the separate parts, the concept looks very promising to deal with the complex and diverse challenges that are ahead of us and ask for an answer.

6.3.4 A First Application of the Water Cycle Concept: Waternet

Till 2006, two public services were responsible for the water management in and around Amsterdam. On behalf of the city of Amsterdam, the public service "WLB" was responsible for drinking water supply. On behalf of the water board Amstel, Gooi, and Vecht (AGV), the public service "DWR" was responsible for wastewater treatment and surface water management, and for the city of Amsterdam for

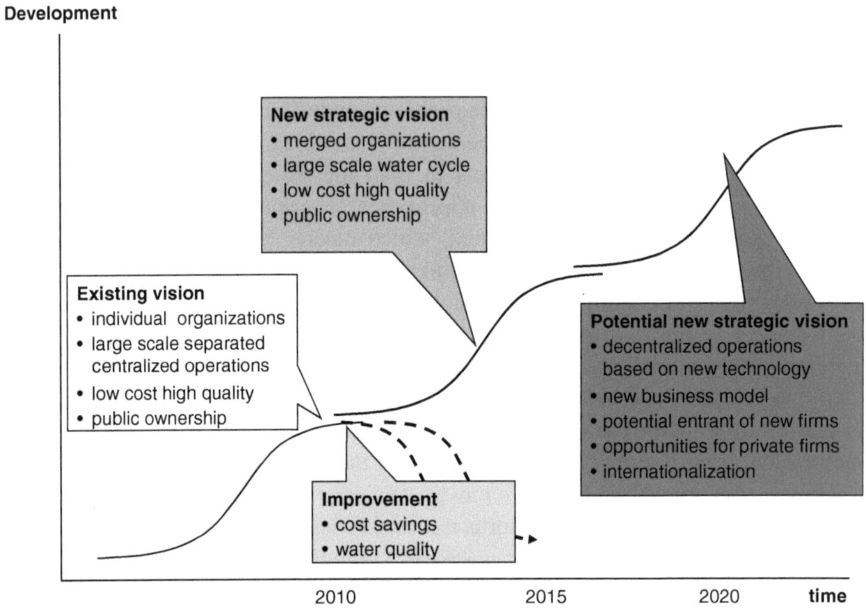

Fig. 6.3 New Strategic Vision Waternet. (Source: based upon T.C.R. van Someren, 2005, Strategische Innovationen, Gabler Verlag)

sewerage. In 2005, both these authorities (city of Amsterdam and water board AGV) realized that, to make the step forward to a company based on the water cycle concept, a merger of these two public services was required. On 1 January 2006, they established the public service Waternet, a merger between DWR and WLB. By merging into Waternet, a new corporation had been created for which a new business model had to be created. A business model describes how activities contribute to the realization of the strategic goals of an organization (van Someren and van Someren-Wang 2006; van Someren 2005). Both increasing bottlenecks in the existing organization as well as emerging trends force organizations to rethink and renew the business model (see Fig. 6.3).

Figure 6.3 shows the development of the business model of Waternet and its predecessors over time. The predecessors of Waternet, DWR and WLB, had reached the maturity phase. In this maturity phase, progress can only be realized by incremental change and gradual improvements. By creating Waternet on 1 January 2006, a jump was made to a new business model, the water cycle concept, characterized by sustainability and integration. This new business model creates new opportunities and ways to reach the strategic goals of Waternet: a systematic and integrated approach of the water cycle, higher quality, higher efficiency, and more customer orientation. At the starting phase of this new business model, radical innovations are required instead of incremental changes and gradual improvements to develop Waternet further along this curve. On the long term, also this business model will reach its maturity phase, and a new business model will be required to survive.

6.3.5 The First Results of Waternet and Future Goals

Waternet integrates all water services in one organization. By combining water services, the expectation is that synergy advantages and improvements can be reached, resulting in a higher efficiency, more customer orientation and a higher quality. In the first 4 years of existence, 2006–2009, the results of Waternet are promising: an efficiency improvement has been reached of euro seven million, the percentage of customers satisfied with the services Waternet provides has grown slightly from 73 % to 75 %, while the total carbon footprint (inside and outside Amsterdam) has been reduced by almost 50 % related to 1990, as an indicator for quality improvement. The long-term aims of Waternet are an efficiency improvement of euro 16 million, a percentage of 81 % of customers satisfied with the services Waternet provides, and a climate neutral operation. Hence, further progress has to be realized.

6.3.6 Innovations to Develop the Water Cycle Concept at Waternet

In order to realize the progress and to establish the long-term goals, Waternet focuses on innovations the coming years. The innovations are based on an integrated and durable approach of the water cycle, beyond the subsystems of drinking water supply, sewerage, and wastewater treatment. Six strategic innovation areas have been identified which are crucial for developing the water cycle concept of Waternet: sources and water system, closing the water cycle, sustainability, new sanitation concepts, new products and services, and innovation in management concepts. As can be expected, these strategic innovation areas are located there where there is overlap, synergy, and interaction between the separate parts of the water cycle. This will be shown in the next paragraphs.

6.3.6.1 Sources and Water System

The strategic innovation area "Sources and water system" is a clear example of connecting drinking water with the water system in the water cycle. Projects, aiming at a strategic use of drinking water sources and the realization of a sustainable and robust management of the water system, will result in an enhanced protection policy of drinking water sources. By taking into account the whole water cycle, the most effective approach to deal with emerging substances can be chosen, resulting in healthy drinking water and ecologically rich surface waters in line with the objectives of the Water Framework Directive. Another chance, especially in the western part of the Netherlands and thus in the area of Waternet, is the use of excess of brackish seepage water as source for drinking water. Brackish seepage water in low-lying polders will increase due to sea level rise and land subsidence, leading to reduced agricultural productivity (Delta Commissie 2008). Abstracting this seepage may prevent salinization of the superficial groundwater while the abstracted water can be used as a new drinking water source. In that way, a "surface water problem" is converted into a "drinking water possibility": the seepage water is free of anthropogenic substances and offers a climate-proof source for drinking water. The first research into the use of brackish groundwater, comparable to the use of brackish seepage water, has recently been started in the Netherlands (Oosterhof et al. 2009). Waternet intends to start a research into abstraction and treatment of brackish seepage water in 2011.

6.3.6.2 Closing the Water Cycle

Although related to the first area, the strategic innovation area "Closing the water cycle" goes further. An example of a technological project is the reuse of treated wastewater for drinking water purposes. At present, there is no real need for this, neither in the Netherlands nor in Amsterdam and surroundings. There are plenty of more reliable drinking water sources, and from the point of risk management, extensive treatment will be required. However, due to climate change, increased variations in river discharges will lead to periodic water shortages and deterioration of the water quality, and thus may change the situation (Zwolsman et al. 2007). Technological breakthroughs by temporarily storing water and desalination including concentrate treatment can make water reuse for drinking water purposes very attractive as it offers a more constant source, both in quality and quantity. For the same reason, closing the water cycle and introducing sustainable water management concepts can be very attractive for industrial applications and industrial water management too. One of the challenges is to develop sustainable technologies that can produce industrial water or process water from the effluent of wastewater treatment plants in a competitive way.

Closing the water cycle can also be applied in the urban water management. An example in this field is the use of treated wastewater for flushing the water system in the town of Amstelveen, located in the service area of Waternet. This has been tested in a demonstration scale project in 2010 and 2011. The treated wastewater will be polished in a constructed wetland before further use. The goal is to improve the water quality in the ditches, ponds, and canals of Amstelveen and to prevent algal blooms. An important research question that will be addressed is the removal and accumulation of hormones and pharmaceuticals that may be present in the treated wastewater.

Another example concerns integrated modeling. Models for drinking water production, drinking water distribution, sewerage, wastewater treatment, and surface water management already exist. Integrating them into one model and incorporating qualitative aspects in these mostly quantitative models will be very useful to optimize the water cycle and to tune all processes. Such integrated models can be used for real time control of the water cycle and process automation.

Closing the water cycle also covers combining tasks and integrating tasks on a managerial level and operational level, taking into account separated bodies. Waternet takes care of the sewerage of the city of Amsterdam. However, in total, 24 municipalities are located in the service area of Waternet. All these municipalities have their own task and responsibility for sewerage, while Waternet is in charge of the entire wastewater treatment in the service area. So, an institutional, partly nontechnical innovation for Waternet is to expand the water services by integrating the management and operation of the sewage systems by these 24 municipalities with the management and operation of the wastewater treatment which is already a task of Waternet. By doing so, the relationship and interaction between sewerage and wastewater treatment are optimized.

6.3.6.3 Sustainability

The innovation area "Sustainability" especially focuses on the development of an integrated climate change policy and climate change actions. Climate change asks for adaptation measures and mitigation measures. Adaptation deals with tailoring the water cycle to a changing climate. Mitigation aims at the reduction of the emission of greenhouse gases. Concerning adaptation, the goal of Waternet is to be climate-proof in 2015. Concerning mitigation, the goal of Waternet is to reduce the emission of greenhouse gases with 50 % in 2010 related to 1990 and to be climate neutral (zero discharge of greenhouse gasses) in 2020. During the International Scientific Congress Climate Change: Global Risks, Challenges and Decisions in Copenhagen, 10–12 March 2009, it became clear that the worst-case IPCC (Intergovernmental Panel on Climate Change) scenario trajectories are being realized, and reductions of greenhouse gas emissions up to 80–90 % are necessary to avoid temperature rises above 2°C.

To fulfill the Waternet ambition, Waternet has developed a climate change action program with 31 actions that cover the whole water cycle. As an adaptation example, research is carried out into the use of new drinking water resources which are climate-proof and more sustainable. As a mitigation example, research is carried out to upgrade biogas from wastewater treatment to natural gas and supplying it to the gas network. In this way, a much more efficient use of biogas is possible, resulting to less use of fossil fuel and less greenhouse gas emissions. Recently, this has been put into practice at the wastewater treatment plant "De RondeVenen" operated by Waternet (Van der Hoek 2010). At this relatively small plant (1,370 m^3/h wastewater), biogas from the sludge digestion (700 Nm^3/day) is converted into natural gas or "Green Gas" (140,000 Nm^3/year) and fed to the gas network. This energy recovery from wastewater is enough to supply 90 households and avoids the emission of greenhouse gases (250 ton CO_2-eq/year).

Another important aspect with respect to sustainability is how to quantify it. Urban sustainability indicators, also for services such as energy and water systems, play an important role in helping policy makers (Keirsted and Leach 2008). Waternet is implementing tools such as life cycle assessment and life cycle costing, and is looking for alternative tools more fitted to the water cycle.

6.3.6.4 New Sanitation Concepts

At present, drinking water supply and sanitation are mainly based on a central treatment. The strategic innovation area "New sanitation concepts" offers opportunities to explore new concepts based on source separation, necessary to make the jump to a new business model in the far future (Fig. 6.3). Exploring new technologies, using nanotechnology and biotechnology, may offer new possibilities for local concepts and new systems. Source separation offers new possibilities for decentralized sanitation concepts. Separate collection of urine looks very attractive to recover P- and N-compounds in an economic way, and is promising for the removal of pharmaceuticals to avoid pollution of surface water and drinking water with these compounds (Guest et al. 2009; Larsen et al. 2009). Separated sanitation also gives good opportunities for energy production, for example, by anaerobic digestion and biogas production (Zeeman et al. 2008; Otterpohl 2002). Recently, a study has been conducted how the water cycle may look like in 2050 for individual houses, new housing estates, renovated housing estates, and rural areas (Ministry VROM 2009b). Different future scenarios have been developed in this study in which all these possibilities and chances of new sanitation concepts were taken into account. New sanitation concepts on the other hand will affect the drinking water demand thus an integrated approach, considering the whole water cycle, is required. In new housing estates in and around Amsterdam, Waternet studies the possibilities of these new sanitation concepts and drinking water supply concepts. At the same time, Waternet is "captured" by the existing large-scale infrastructure having a very long depreciation period. The challenge is how to make the transition to new sanitation concepts in a sustainable and economic way.

6.3.6.5 New Products and Services

The creation of a water cycle company results in new business opportunities to realize new services. Especially two new developments are worth mentioning. The first is related to customer satisfaction. Until now, the citizens of Amsterdam are confronted with four tariffs which are each billed separately: a tariff for drinking water, a tariff for sewerage, a tariff for wastewater treatment, and a tariff for water system management. With the creation of a water cycle company, it is possible to put all these tariffs on one combined bill. For a customer, this has the advantage that with such a combined bill, showing all water services clearly specified, he can see at a single glance what he has to pay in total for all water services, enhancing transparency. A further step is to combine the drinking water tariff, the sewerage tariff, and the wastewater treatment tariff into one combined tariff. This may encourage the customer to a more sustainable use of water.

A second development concerns the relation between energy and water. Water can be used as an energy carrier, but water also contains energy itself, either chemical energy (organic compounds present in the water) or thermal energy (heat and cold). A water cycle company manages drinking water, wastewater, surface water, and groundwater. All these "water types" offer possibilities to use the energy content of the water. From (too) warm drinking water, heat can be recovered. With the very good isolation of new houses, most of the energy leaves these houses with the wastewater. Extracting the thermal energy from the wastewater can be very attractive. Extraction of deep water from lakes can be used for cooling purposes. Finally, groundwater can be used for the so-called Aquifer Thermal Energy Storage (ATES).

Drinking water treatment, sewerage, and wastewater treatment result in energy consumption. On the other hand, energy can be recovered from wastewater by anaerobic sludge digestion, resulting in biogas. Thermal energy can be recovered from the wastewater by using heat exchangers. In 2007, the total energy consumption is 5,580 TJ while only a small part of the chemical energy is recovered from the wastewater. In 2007, the energy balance is negative. For the future situation, it is assumed that all the energy from the wastewater is recovered, both chemical energy and thermal energy. This results in a positive energy balance. Focusing only on wastewater as energy source, it can be concluded that already with this source the water cycle can be transformed from an energy-consuming to an energy-producing cycle. Taking into account the energy content of the other types of water, the potential is even much higher.

Energy from the water cycle can be used to operate the water cycle climate neutral in the future. In 2009, the purchase and use of renewable energy avoids the emission of greenhouse gases, but still there is a residual CO_2-emission of 62,800 tons which has to be compensated. For a part this compensation can be realized by energy savings, and for a part it can be realized by production of renewable energy. Of course, wind energy and solar energy are possibilities, but for a water cycle company, it is much more attractive to use the energy content of the water. As already mentioned, there are a lot of possibilities for energy recovery from the water cycle. The challenge is to find the most robust combination of energy sources able to fulfill the demand at any time. Smart combinations are necessary to abandon the use of fossil fuel and become climate neutral.

A more classical combination of energy and water is already put into practice in Amsterdam. The wastewater treatment plant is located near the waste incineration plant. Sludge and biogas from the wastewater treatment plant are used as fuel for the incineration plant. The heat and electricity produced by the latter are in turn used for wastewater treatment by Waternet. By doing so, the use of 1 million Nm^3 of natural gas per year in this wastewater treatment plant is avoided which equals an avoided greenhouse gas emission of 1.8 kton CO_2-eq/year. A combination of water and energy thus leads to sustainable solutions that under the right circumstances can lead to an energy-neutral or even an energy-producing water cycle (Kiestra 2008).

6.3.6.6 Innovations in Management Concepts

The strategic innovation area "Innovations in management concepts" deals with innovations in the field of management. Important is to realize that the three institutional players in the water cycle have a strong focus on management and control. The concept of the water cycle however asks for change and innovation. That is why innovation management at the water cycle company Waternet is so important. In innovation management, the main issue is to create and stimulate innovation and creativity on the one hand and to guide innovation in a preferred direction on the other hand. It is the delicate balance between bottom-up and top-down, between room for free experimental work, and strict, clear targets with performance indicators.

Secondly, alliance management is important. To realize innovations, Waternet as a water cycle company has to work together with knowledge partners and suppliers. How to choose the right partners, how to manage the cooperation, and how to manage the alliances are the key issues.

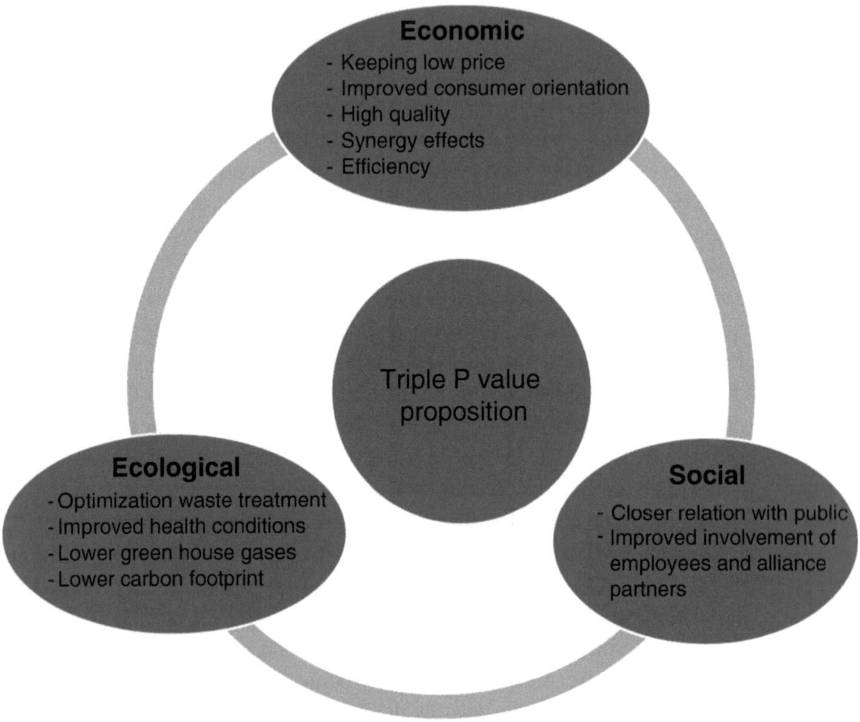

Fig. 6.4 Triple P value proposition Waternet

Originating from predecessors with a clear task, innovation management and alliance management imply new management concepts for a water cycle company.

Waternet has created possibilities for innovations in the water cycle in which employees with a different background in drinking water, wastewater, or water system collaborate and find new solutions. Financial resources and human resources have been allocated for this purpose. Also new alliances have been established with new partners, focusing on the water cycle and innovations in the water cycle, among others with Delft University of Technology, Wageningen University and Research Center, the University of Amsterdam, KWR Watercycle Research Institute, and with water companies like Hamburg Wasser and Singapore Public Utility Board.

When summed up, the Triple P results are shown in Fig. 6.4.

The Triple P of the water cycle company is characterized by improvements in ecological and economical efficiencies which would not be achievable with the separate organizations and their traditional business models. However, the potential third growth curve of decentralized drinking water company shows the potential of another radical improvement of Triple P indicators.

6.3.7 The Future of the Water Cycle: Critical Success Factors

The water cycle is an encouraging vision. The above mentioned strategic innovation areas open several new options for the strategic development of the water cycle in general and for Waternet as the first water cycle company in the Netherlands in specific. As shown in Fig. 6.3, different growth curves based on a different combination of strategic innovations can be created.

The further development and growth of the water cycle concept requires a tremendous effort of all parties involved in order to align all necessary technical, organizational, and administrative changes.

Three factors, based on "The Necessary Revolution" (Senge et al. 2008), are the pillars to make the transition from a fragmented water cycle to an integrated, holistic water cycle successfully. Firstly, the developments in technology and society require that we see the water cycle as a system in which the three domains drinking water, wastewater, and water system interact. Especially at the interfaces, new opportunities appear, and connections between the separate fields of knowledge create new solutions. Secondly, only by collaboration across boundaries, we can take full advantage of the water cycle concept. We have to build bridges between the separate elements of the water cycle and look for the cross connections. Finally, we have to be creative to find the best solutions to make the transition to a sustainable and "future-proof" water cycle. Strategic innovation is an important key for this transition.

In the recent years, the international water market has changed dramatically. Water and water management are being regarded to be core challenges for societal development. The water market has become increasingly commercial with great opportunities for private and public organizations. The Chinese market is a multi-billion market and is still increasing. Water-related companies of France, Germany, and the USA are trying to grasp their share. The solutions offered need to be turnkey, and only commercially integrated firms are able to satisfy this demand. Although the Netherlands possesses a large share of water management knowledge and technology, about 25 % worldwide, the market share in Asia is only 1 % (van Someren and van Someren-Wang 2006). Commercial exploitation of knowledge requires strategic innovation of the total water sector.

Moreover, one of the trends is going into the direction of decentralized concepts requiring new business models. The consequence of decentralization is not only a technical challenge but opens the opportunity for private firms as well. These developments have been discussed with the sector in early 2000, but most managers had the opinion that these signals were too weak to be a real threat to the existing organizations and future sector as a whole.

In Table 6.2, some examples of challenges for the Chinese water industry are presented.

On all levels, macro, meso, and micro, the challenges in the Chinese water sector are huge. Examples are the discrepancy of water supply and demand between North and South China, the infancy of the water management on the regional level, and industrial wastewater treatment. Systems with high reliability and quality have to

Table 6.2 Chinese management agenda for the water industry

Megatrend	Issues	Examples of challenges
Climate change	Value circle	Connection between water systems, industries, and human consumption
Nanotech, biotech	Strategic innovation	New technology fields, focus on nontechnical innovations, decentralized concepts, link between legislation and institutional water management, focus on technology appliance instead of construction, cross-industry innovations
Commercial markets	Business model	Focus on low-cost-efficient processes with high EHQS (environment, health, quality, and safety) standards, highest transparency with regard to operational activities and accidents
Health and environment	Sustainability	Integration of Triple P criteria, legislative execution power, water policies, discharge guidelines, and execution power for industries
Professionalism	Operational	Efficient operation of primary and secondary processes, achieving high reliability standards, accident and crisis management

renew current Chinese water management in order to better cooperate and control industries and deal with accidents and crisis situation. New technologies and in particular technologies outside of the water sector itself such as biotech and nanotech have the potential to revolutionize the water sector. Moreover, the blurring of industry boundaries will characterize the future of the industry. For example, in megacities the decentralization of energy supply fits to the opportunities for decentralized water supply and energy reuse of warm water in households. The soft factors of water management will prove to be the most difficult part for the technology-driven Chinese. Water pollution can have a devastating influence on health, and an adequate accident and crisis management is necessary.

6.3.8 Lessons Learned

Lesson 1: The monopoly of the sewage and drinking-water company did not lead to high prices, high cost levels, or low quality. On the contrary, extreme high quality and high reliability was combined with low m3 costs and low consumer prices. The price of 1,000 l of drinking water is the same as 1 l soft drink. It demonstrates that public-owned organizations are able to deliver high-performance and bottom-line results.

Lesson 2: Due to being a monopoly and a public organization, the reaction time on new trends like decentralized drinking water opportunities is rather low. Consequently, the speed of strategic innovation is rather low. For the public water companies, the rule is quality, safety, and low cost first. Innovation is second. Water companies of the future need a higher response time, and the innovation function has to be institutionalized in order to cope with new challenges and changing market structures, behavior, and desired outcomes.

Lesson 3: Public organizations need strategic innovation as well. It is often thought that public organizations do not need strategic innovation because they are not part of the market. The contrary is true. Because of worldwide trends, changing societies, and so on, every organization including a public organization need strategic innovation to prove their reason to exist and to have an appropriate service with a high quality against low cost.

Lesson 4: The long history of water management and its organizations sets the mindset on improving the existing situation. Thinking out of the box is extremely difficult also due to the success in the past. The larger the success and an acceptable status quo, the more the weak and string signals of changing conditions are ignored or waved away as not relevant yet.

Lesson 5: Organizational innovations pave the way for technical solutions, from value chain to value circle by closing the water cycle. The integrated firm can easier take care of the interfaces between sanitation and drinking water sectors.

Lesson 6: The boundaries between industries are being blurred as is the case for water and energy. The energy component in, for example, warm tube water has a high potential for recovering heat. But also contributions of other industries like biotechnology and nanotechnology for the primary drinking water and sanitation processes are increasingly becoming relevant.

Lesson 7: The fragmented network organization is slowly transformed into integrated firms to reduce complexity and reduce uncertainty. Accumulated knowledge within separate firms leads only to local improvements and hindered regional strategic innovations.

Lesson 8: The increased size of the company by greater economies of scale and scope makes it a more suitable partner for alliances to cope with future challenges. The coordinator role increases speed of stimulating necessary innovations.

Lesson 9: The dominance of the public organizations ignored the trend toward privatized and liberalized international water markets. The possession of knowledge is not commercialized. It proves to be extremely difficult to accept other rules in other markets. This contra innovative attitude hinders the commercial appliance of available knowledge.

6.4 Agro-food Sector

The cases discussed here are based on a study by a Dutch innovation agency for the agro-food sector (van Someren and Nijhof 2010). The aim was to find out why Triple P innovations in the agro-food sector are hard to find, and why a business wise approach is often largely lacking. The strategic innovation theory was used as a foundation for analyzing the cases.

Like the water sector, the agro-food will be a crucial industry for the future due to the supply of food of an increasing world population and the accompanying sustainable effects. In the last decades, with the increasing urbanization, the rural areas and the cities are more or less separated from each other. Phenomena like city marketing, cultural sector, and attracting creative industries to revive city economies took the attention of many Western policy makers. European agricultural policies were supply-driven, economies of scale, and heavily subsidized due to political motives (votes and sustaining traditional businesses). The historical band and economical connection between city and rural areas were cut with the consequently global value food chains. Agriculture is old-fashioned business and causing problems with all kinds of diseases. The intensive livestock industry is linked with diseases like avian flu, Creutzfeld-Jacob, and swine influenza. Mass poultry farms and intensive agriculture lead to the use of fertilizers and corresponding detrimental effects for, for example, groundwater. In fact, the agro-food sector is for many parties based on mass production economically very successful but not very sustainable when ecological and social effects are taken into account.

At first sight, it is not a very attractive industry to work for. The contrary is the case due to new opportunities based on new technologies and organization of global food chains. Fundamental rethinking about our food supplies is necessary, and a complete inversion from large-scale unsustainable supply-driven production to market-driven sustainable business propositions is necessary. In our research based on nine cases in the agro-food sector, revolutionary changes are possible (van Someren and Nijhof 2010). Examples are sustainable fruit production for which a complete new value chain has been set up, a circular business model by means of connecting the CO_2 and heat waste stream of a large fertilizer plant to glass house sector, and a model for establishing a greenport near to Shanghai. All cases revealed that individual entrepreneurs created new value chains in which Triple P was the bottom line. But all the initiatives are on a relatively small scale, and the next challenge is to grow the activities and become a threat for incumbents dominating the agro-food trade.

6.4.1 Metropolitan Agro-food

The case of metropolitan food is a collection of different projects aiming at making the agro-food sector more sustainable. In our assignment, ten outstanding cases have been evaluated on their contribution to Triple value and new business development.[5] The history of agriculture may be as long as the history of mankind. In all these centuries, agriculture has changed sometimes dramatically due to the introduction of new tools, new species of plants from other parts of the world, artificial fertilizers, and mechanization as the major landmarks. More recently, after the Second World War, major landmarks in the agro-food sector from an innovative view were the introduction of factory farming and the green revolution based on a combination of plant breeding, irrigation development, and the financing of agrochemicals.

[5] This section is derived from T.C.R. van Someren & A. Nijhof, 2010, Triple P Business Development in the Dutch agro-food sector, van Gorcum Publishers.

Over the years, our perception of farming also changed significantly. It once started as a human activity which yields valuable products that mankind needs for its survival, but nowadays our perception of farming is much more fragmented and dispersed. It is also perceived as a sector related to many issues of pollution like contamination of groundwater with toxic pesticides, high concentrations of phosphorus in soil due to overfertilization, and as a major contributor to global warming due to the greenhouse gases related to livestock.

Next to this, farming is also heavily involved in the debate about the use of scarce resources; think, for example, of regions with water shortages and the effects of farming on rain forest preservation. It is also related to issues like the affordability of food in developing countries and the diversion of food crops to crops used for biofuels. Another issue that is especially relevant to the cattle breeding sector is the linkage of farming to animal welfare and human health. Especially in regions with an intensive livestock industry, nongovernmental organizations emerged to protect animal rights.

This debate was intensified at the beginning of the twenty-first century due to several global diseases like the avian flu, Creutzfeld-Jacob disease, Australian Q-fever, and swine influenza. This fragmented and dispersed landscape with respect to agricultural production puts forward a major challenge for the future. Current production has to take into account the social and ecological issues it runs into. Sustainable development requires thinking and acting from a Triple P, bottom-line perspective, taking into account economic prosperity (profit), environmental quality (planet), and social justice (people).

However, the challenge for the future is more demanding. The need for agricultural products is increasing rapidly. Due to the growth of the world population and changes in consumption patterns because of higher welfare levels, agricultural production must expand by 70 % until 2050. For this it is not enough to incrementally improve the current way of production, because it is the current way of production that caused many sustainability issues. Instead, sustainable development in the agro-food sector requires innovation toward new ways of production and consumption. This goes well beyond normal interaction in command-and-control policies. Rethinking current business models, objectives of value creation, and new terms of engagement seem to be first-line answers. There are some basic strategic lines.

One direction for innovation in the agro-food sector is focusing on combining farming activities with other activities. In 2007, one out of ten farmers had developed some additional activities to increase the income of their company. An illustration is Green Care in which some farmers included care activities by taking up handicapped people or people who need psychiatric support. These activities are grouped together under the label "multifunctional farming." A second direction for innovation in the agro-food sector is focusing on making the current farming activities more sustainable. This results in initiatives to develop new stables that combine improved animal welfare with efficient production methods, in initiatives where landscape design is integrated with current agricultural activities, or in initiatives where new forms of agricultural production are introduced.

A third direction focuses on the linkages within the supply chain of certain agro-food products. Direct supply from producers to retailers of fruits and vegetables can result in shortening of the supply chain, less environmental impact, and a fair distribution of financial margins. Other initiatives focus on broadening supply chains by using the waste of one part of the agro-food sector as input for other parts. Examples are the development of biofuel plants and linking agro-streams to biomass fermentation in Biopark Terneuzen. The creation of an agro-food park based on a value circle shows Greenport Shanghai (van Someren and Nijhof 2010).

6.4.2 Green Egg

For centuries, an egg was an egg. Since the introduction of the Rondeel egg, it is not anymore like that. Over the centuries, new methods for keeping hens were introduced, which resulted in huge efficiency improvements. Especially after the Second World War, the use of cages became dominant in this sector, facilitating extensive and continuous volumes of egg production. Furthermore, specialized organizations were introduced in the supply chain of egg production and consumption. These arranged food for the hens, the packaging and logistics of eggs, and all retail activities for selling the eggs to the consumers.

However, in developed countries dissatisfaction with the production of cage eggs has increased significantly over the last years. As a result, the European Commission set new norms for animal welfare prohibiting cages in 2012 (nota 1999/74/EG of the European Union). In the meantime, most of the supermarkets in the Netherlands already stopped selling cage eggs due to protests of nongovernmental groups like "Dierenbescherming" (society for the prevention of cruelty to animals) and "Stichting Wakker Dier" (Alert Animals Foundation). This resulted in the gradual introduction of improvements in production methods, where hens have more space to move about ("barn eggs" and "free-range eggs") and are fed organic food ("organic eggs"). These improvements can be characterized as incremental improvements because certain parts of the production or consumption chain are improved without rethinking the whole supply chain.

The Rondeel case has a different ambition and aims for radical improvements. It strives to produce sustainable eggs and takes into account the whole range of social, ecological, and economic issues related to the production and consumption of eggs. The start of the Rondeel case can be traced back to a crisis situation. In 2003, many countries were confronted with the avian flu, which brought to light the contamination risks of free-range eggs and intensive hen farming. It was the Dutch minister for Agriculture, Nature and Food Quality, Cees Veerman, who instigated for a project to rethink hen housing.

The case of Rondeel egg demonstrates the design and implementation of a new barn and related supply chain for producing eggs under improved conditions of animal welfare, improved environmental conditions, and sound economic returns (Fig. 6.5).

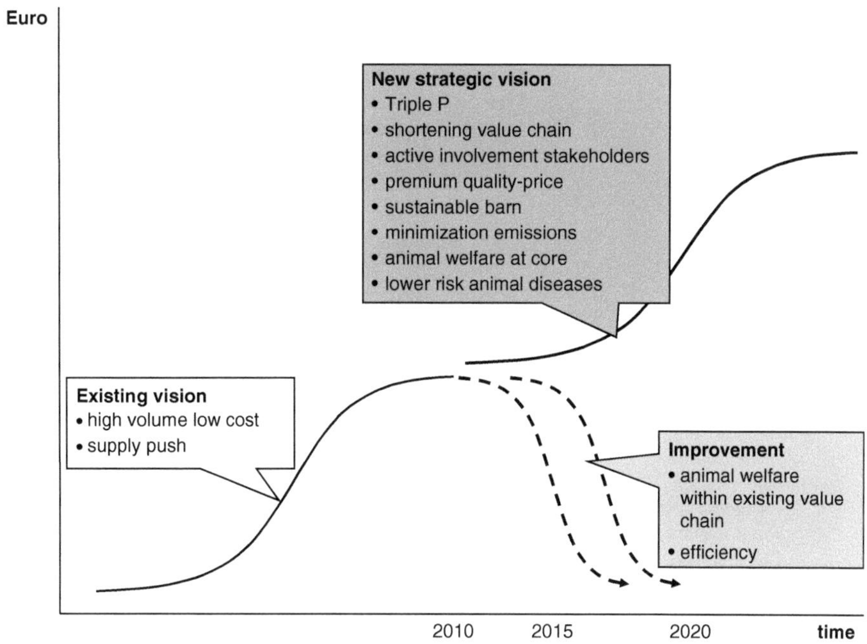

Fig. 6.5 New strategic vision Rondeel. (Source: based upon T.C.R. van Someren, 2005, Strategische Innovationen, Gabler Verlag, van Someren & Nijhof, 2010)

The value proposition of Rondeel started with talks during the project "hen keeping." This project focused on three different questions: "What do civilians want from an egg?" "What does the chicken want?" and "What does the poultry farmer want?" These guiding questions were meant to find the starting points for a truly sustainable egg. Based on talks with representatives of all interest groups, the University of Wageningen drew up a list of values that had to be taken into account for a sustainable egg[6]: hen welfare, ecological sustainability, transparency of the supply chain, fit the landscape, higher returns for the farmer, innovation and exemplary function, and optimal coherence between all parts of the value chain.

One Rondeel barn consists of 5 day/night compartments for 6,000 hens each. Therefore, the total number of hens in a Rondeel barn is 30,000. For 2010, it is estimated to build one new Rondeel barn in the Netherlands every 4 months. Furthermore, there are talks going on in Belgium, the UK, and Austria to invest in similar barns.

With a weekly production of 150,000 eggs, one Rondeel barn produces about 0.08 % of the total egg production in the Netherlands. If all eggs in the Netherlands would be produced in Rondeel barns, this would allow for 1,200 Rondeel barns. There are clear examples of value conversion in the business model of Rondeel. The Rondeel egg requires a surplus price in the market that is particularly based on improved animal welfare. This claim is backed up by a study of Wageningen

[6] Page 10 of the end report phase 1 of "Op zoeknaar het Gulden ei," January 2009.

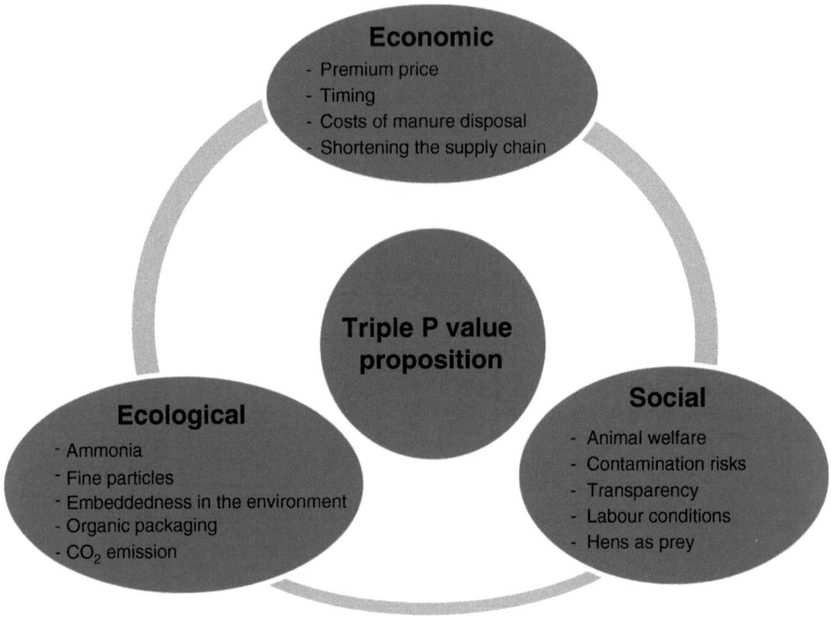

Fig. 6.6 Triple P of Golden Egg

University (van Someren and Nijhof 2010) stating that on a 10-point scale the enriched cage system results in a score of 2.3, a barn system with a covered free range in a score of 6.3, a free-range barn or aviary in a score of 7.8 and the Rondeel barn system in a score of 9.6. This is even above the score of uncultivated poultry where hens roam free (score of 8.7). The good conditions with regard to animal welfare are also put forward by the Dutch nongovernmental organization "Dierenbescherming" that awarded the maximum three-star rating to Rondeel eggs, the same rating as organic eggs. The Rondeel egg is also the only egg with a "milieukeur," a trademark indicating relatively low levels of energy use and environmental burden.

In comparison to the value proposition of the traditional egg, the Rondeel egg has many different aspects in relation to the Triple P concept. In Fig. 6.6, the Triple P aspects of the new value chain for the Rondeel egg are summarized.

The radical rethinking resulted in a business idea where all aspects of the Triple P are an integral and inseparable part of the Rondeel egg. The Triple P outcome is dependent on several players in the value chain. The supermarkets are the powerful player, and they determine consumer pricing and marketing leading to the scale of consumer acceptance and buying behavior. The innovative barn concept and the strategic innovations like the involvement of the Dierenbescherming and their three-star rating fulfill the necessary conditions for success.

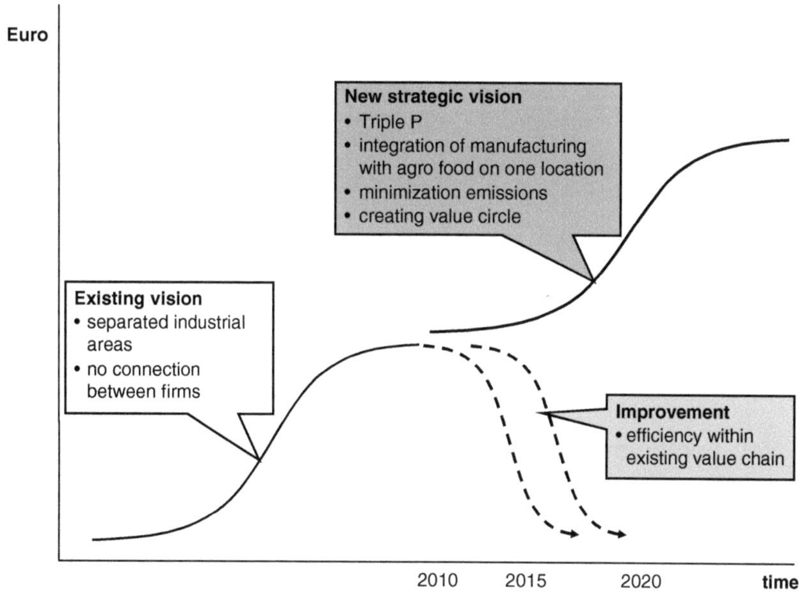

Fig. 6.7 Biopark Terneuzen. (Source: based upon T.C.R. van Someren, 2005, Strategische Innovationen, Gabler Verlag, van Someren & Nijhof, 2010)

6.4.3 BioparkTerneuzen and Greenport Shanghai

6.4.3.1 BioparkTerneuzen

The Dutch greenhouses are worldwide famous for their economic efficiency and high output of quality products. In several different geographical areas throughout the Netherlands, the greenhouses were clustered. Besides economic efficiency, also ecological efficiency got attention of the farmers due to high energy prices. New technical solutions for the energy installations were tried out such as heat-power combinations. In Biopark Terneuzen, a new strategic innovation was introduced as Fig. 6.7 shows.

In Biopark Terneuzen, it was tried to develop an agro-park based on the introduction of biofuel plants and the reuse of waste streams like heat and CO_2 for greenhouse farming. For the first time, the greenhouses were not located in traditional agro-food farm areas but in a capital intensive manufacturing area. One of the largest manufacturers was a Finnish-owned fertilizer company with large CO_2 emissions. The idea was to close the cycle between two different industries by coupling the output of CO_2 from the fertilizer factory as input for the greenhouse companies. In this industrial area, also other companies like the biofuel company were located. The closing of the cycle with regard to the secondary processes was expected to lower costs with a reduction of environmental detrimental effects. However, in practice it came out that the desire of the farmers was to be independent in order to lower the risk of disruption in energy supply. Companies made

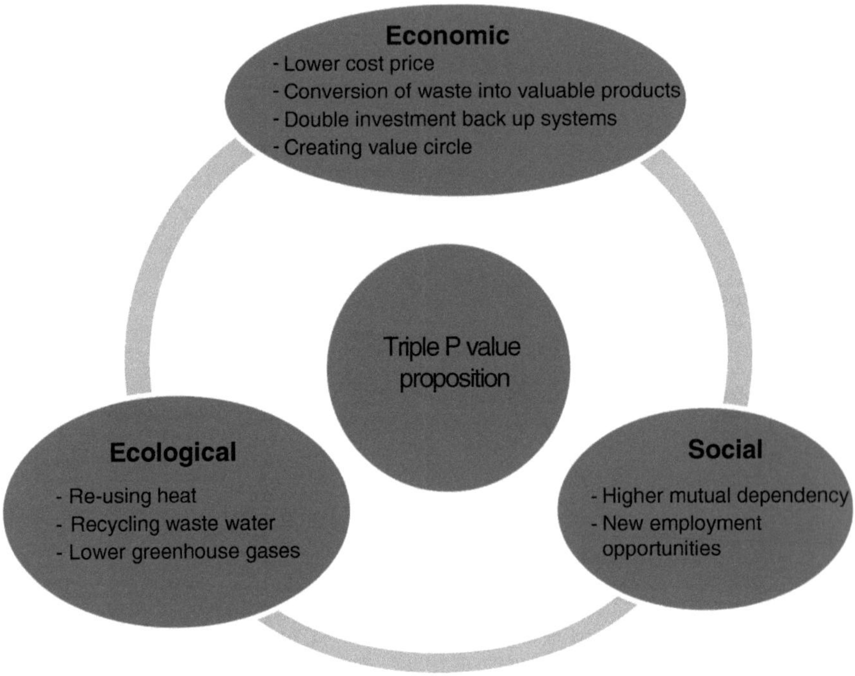

Fig. 6.8 Triple P value conversion Biopark Terneuzen

double investments to be secure of energy supply, and back up installations were installed. The risk of loss of revenues from the primary process was higher than saving on costs in the secondary process. The mutual dependency resulted in higher costs and investments (Fig. 6.8).

The Triple P of the Biopark shows the interdependencies of the three elements. The physical secondary flows can be closed and forming a value circle resulting in lower costs and lower emissions by a given output. But there is a price to be paid. The higher mutual dependency leads to double investments in order to lower risk of breakdown of heat and CO_2 supplies. It is also a part of the culture of greenhouse farmers to work as a free entrepreneur and independently of any other. In a biopark, exactly this cultural entrepreneurial feature is taken away. The Biopark Terneuzen case demonstrates the difficulties in the attempt to create a value circle in which economic aspects appeared to be pivotal for success.

6.4.3.2 Greenport Shanghai

There is another attempt to create a value circle with totally different visions and aims, showing some other difficulties. It is the case of Greenport Shanghai. Here, we only give some highlights demonstrating a complete different situation.

In the case of Greenport Shanghai, plans by several Dutch and Chinese parties were made to develop an agro-park around the metropolitan city of Shanghai based on transferring Dutch knowledge and experience in agro-parks to a Chinese setting.

Table 6.3 Chinese management agenda for the agro-food industry

Megatrend	Issues	Examples of challenges
Metropolitan agro-food	Value circle	New connection between (mega) city and rural areas, blurring of industries
Food shortages and bio-based fuels, biotech, gentech	Strategic innovation	New technology fields, focus on nontechnical innovations, synthesis of food production, and bio-based fuel production
Mass production	Business model	Transformation from small-scale production to professional sustainable high-quality mass production, creation of new value circles based on sustainable food products
Food safety, health, and environment	Sustainability	Integration of Triple P criteria including animal welfare, legislative execution power, internal management systems focusing on quality and food safety as first priority
Professionalism	Operational	Training and professional management

The concept is based on local food production thereby concentrating business in the agro-food park and reducing ecological detrimental effects due to shorter transport distance and closing material cycles within the park, leading to a minimum of waste and emissions. The social effect was to involve local farmers for future production and to increase the level of food safety.

The designed plan was a theoretical sound, but it went wrong in the roll out and commercialization phases. Not only the Dutch side lacked large investors prepared to put money in the project but also a wrong approach to deal with the Chinese caused a no go after many negotiations. The master plan for the agro-food park was much more a technical master plan than a business plan. Wrong assumptions on markets, energy prices, and waste were assumed leading to a wrong business picture. In fact, based on the technical plan, there was not a business case leading to value conversion. The economic dimension of the Triple P turned out to deliver a negative contribution.

Furthermore, wrong parties were chosen and involved to carry out the business development. This cross-cultural factor in alliance management was a false start and neglected from the beginning, although early warnings had been given. However, although this attempt was courageous, it was also naïve, but the basic idea is not wrong. When both new realistic assumptions are applied in new circumstances, as well as the proper attitude and approach with dealing with Chinese is applied, there are still opportunities to realize green agro-food parks. Starting never seen agro-food parks in a distant country like China may be a step too far.

These cases from the agro-food sector demonstrate the broad scope of strategic innovation potential. For Green China, a few typical challenges are summarized in Table 6.3.

Green China will make use of new technologies currently developed in the biotech and gentech industries. The demand for food exceeds the current supply capacity and is a huge sword of Damocles for Chinese government. The potential social unrest due to food shortages and food safety problems is a danger for any government. A fundamental redefinition of the coexistence and boundaries between city and rural areas leads to a restructuring of supply demand coordination. The rise of the megacities and higher income change the quantity and quality aspects of the agro-food industry. Also here, crossover opportunities between industries potential with renewable energy, bio-, and gentech require strategic innovations. For example, some gen-manipulated rice experiments reduce the water consumption of growing the rice by half. Furthermore, the demand for food and fuel challenges both the size of the available agricultural land and available crops.

6.4.3.3 Lessons Learned

Lesson 1: Several trends such as increasing urbanization and increasing demand for safe and ecological viable food create the basis for strategic innovation in the agro-food sector. It appeared to be difficult to fight against greenwashing and to implement objective claims of ecological progress. The Triple P yardstick lacks a clear transparent definition or standard and subjective trade-offs will be inherent to its use. Hence, room for shifting criteria exists, and the line between true green and greenwashing is very thin. Nevertheless, just because of failing transparency, new opportunities for creative entrepreneurs exist to experiment and create new sustainable value chains and circles.

Lesson 2: Organizational–spatial innovation such as metropolitan food and coupling of manufacturing with food production form the basis for a new potential new agricultural revolution. Therefore, the creation of value circles has consequences for the current ruling spatial–organizational–institutional–legislative framework. Implementation requires the buildup of a new framework and the (partial) breakdown of the old one.

Lesson 3: Ecological aspects may be the trigger for change, but economic dimensions decide about success or failure. These economic effects often go beyond individual parties. As soon as solutions are implemented within the existing chain, it depends on the interests of parties to participate. For this reason, the creation of complete new value chains with new parties or parties prepared to change their operations is a long way but probably faster.

Lesson4: Success of master plans for large-scale strategic innovation is dependent on individual decisions and interests of involved parties and current operations. The technical concepts could be developed by creative small organizations, but the large-scale master plans require large-scale parties and large-scale financing for implementation. Doing business with the Chinese requires this large-scale approach.

Lesson 5: In the end, the cultural differences between Western and Chinese parties appeared to be more important than the technical feasibility of the master plan.

6.4.3.4 Ten Golden Lessons

What lessons can be learned from these cases. Of course, three cases are not statistically relevant for law-like lessons, but from our practice, we know that other cases show similar developments and behavior of participants.

Lesson 1: Green China Needs a Strategic Innovation

For Western-existing corporations, the biggest lesson is to pay more attention to the exploration side and organize the institutionalized innovation system accordingly. The challenge is to bridge the jump from the old growth curve to the new growth curve.

For Chinese organizations skipping the growth curve of old technology, it is essential to have an institutional innovation system supporting the commercial development. The advantage of an active and participative government can make the difference by creating the necessary infrastructure. The latter will not be the problem, but the institutionalized innovation system within and between organizations is the core issue to be solved.

The success of Chinese organizations is to couple existing proven technology with demand for high quality and low costs. In order to survive these, relatively new companies need to be able to adapt themselves to new circumstances. The need for strategic innovation capacity will increase with the rise and growth of industries and firms along the current growth cycle. Future sustainable China will become more dependent on strategic innovation and less on green technology.

Lesson 2: Strategic Innovation Is Not Easy but Can Be Generated

The involvement of employees and other stakeholders in the bottom-up innovation process contributes to unexpected and rewarding new outcomes and options. The insights, knowledge, experiences from own and other industries, and creative thinking of all people involved lead to a process of unmatched innovative new client solutions and market opportunities. The combined bottom-up and top-down innovation processes require a leadership style which sets requirements for being able to allow for really new thinking, accepting better ideas from others other than the Board, giving away power to lower management levels, governing and leading on both process and content. Western "modern" concepts like co-innovation and open innovation get a complete different meaning in a Chinese context and cannot be copied.

Lesson 3: The Way to the Solution for Green China Needs to Be Found

There is no standard solution. The history, current business operations, experiences, and vision about the future within any organization differ. Hence, each situation requires a unique approach. But the good news is that the strategic innovation theory (SIT), discussed in Chap. 5, offers the methodology and working procedures to cope with this challenge. Every organization has to find and define its own path in creating the future. Contrary to the opinion of many gurus, there exist no recipes and one-for-all solution. Strategic innovation is a journey featured by blood, sweat, and years of endurance and perseverance. But it is the most rewarding journey of your business life.

Lesson 4: Timing and the Accumulation Effect of Time Is Extremely Important for Launching Strategic Innovations

Being too early is a waste of money and other resources, and being too late leads to unrecoverable lost profits. The right timing is the most difficult entrepreneurial decision. In any strategic innovation, the accumulation effect of knowledge, experiences, business relations, and competences is almost every time underestimated. The reason is that, at the beginning of any strategic innovation, the current technology or business model is always better in performance, quality level, cost level, mature proven technology, and customer acceptance. But due to the time factor, in the eyes of existing businesses, the tipping point always comes suddenly and earlier than expected. All the elements of the strategic innovation fit together and from that moment on is very difficult to catch up. For many it is too late, and the newcomers win. Especially when the timing is also right, newcomers can be unbeatable.

Lesson 5: Nontechnical Innovations Are King

Almost any technology can be copied easily and within a relatively short period of time. Therefore, becoming unique and gaining a sustainable advantage on your rivals are dependent on nontechnical innovations. Even more important than focusing on competition is the ability to offer products and services which are accepted by the market, which requires that you are able to reinvent yourself even if you are successful now.

Lesson 6: Green China Needs Long-Term Vision

Managers but also financial stakeholders have a biased preference for short-term profits. Cost-reduction programs contribute directly to profit and loss statements contrary to exploration budgets. This is much easier and less riskier for your own career than starting up a radical new program. Most European financial stakeholders like venture capitalists, and commercial banks prefer short-term results, which is detrimental for creating the right business model and an adequate business development path.

Lesson 7: Brilliance in Technology Now Can Be a Barrier for Strategic Innovation

Focus on both old and new technology leads to a tunnel vision. In the agro-food sector, the focus on technical brilliant master plans is a myopic view in the business world. The hybrid case shows that managers can stick to viable technical arguments against hybrid technology but miss the marketing effect and the reaction of the larger public and politics.

Lesson 8: Success and Failure Is a Paradox Wheel Which Can Turn

Current success is not good for out-of-the box thinking. Having success in current business can close eyes for alternative solutions. Continuing on this track of success will be in the short term profitable, and who of the managers dares to challenge a successful road of the top? Adopting a really strategic innovative path within a successful organization is extremely difficult. The water company showed that changes in many fields are necessary especially in the political arena and their affiliated organizations. In case of the automotive case, the idea to transform to a mobility service provider instead of producing and selling cars was too far out of the box. There you need newcomers who take up this idea and challenge the incumbents.

Lesson 9: Not Only Private Organizations Have Growth Curve

Strategic innovations are relevant for private firms and public firms, MNEs and SMEs, and manufacturing and services. For private organizations, it is almost a no brainer that they need strategic innovations. Much less logical and much less attention was paid to public organizations. Public organizations need strategic innovations even more because they have neglected these renewal processes almost completely. Especially in relation to Green China, truly new thinking, services, and roles played by public organizations are necessary.

Lesson 10: Strategic Innovation Goes Beyond Business Model

Changing the business model needs not only creative thinking about the business model itself but beyond. The selection of business partners, cross-cultural influences, and governmental relations all determine substantially bottom-line results and success rate. All cases showed the need for new competencies needed to make the innovation a success. The upside is that the accumulation can lead to competitive advantages, and the downside is that old competencies and knowledge that laid down corporate experience and workforce can get obsolete.

The lessons demonstrate the need for innovative entrepreneurship and strategic innovation in both the private and public sector. They also show that, due to the features of sustainability depending on the market and specific circumstances, the government has to play a role. This conclusion has also been drawn earlier when the rise of China was discussed. Therefore, it is now time to turn to the contribution and roles of government.

References

van den Berg, G. (2009). Substances of concern for the drinking water supply from the rivers Rhine, Meuse and Drentsche Aa. Report KWR Watercycle Research Institute Nr. 09.012: Nieuwegein, the Netherlands.

de Bruijn, F. A., & van Mazijk, A. (2003). *Effects of climate change on the quality of the river Rhine water*. Report RIWA (Association of the River Waterworks): Nieuwegein, the Netherlands (in Dutch).

Delta Commissie (2008). *Working cooperatively with water*. Report of the Dutch Delta Commission: the Netherlands (in Dutch).

van Dorland, R., Dubelaar-Versluis, W., & Jansen, B. (2009). *The status of the climate in 2008*. De Bilt/Wageningen, the Netherlands: Publication of the PCCC (Platform Communication on Climate Change) (in Dutch).

European Union (2000). Directive 2000/60/EC of the European Parliament and the Council of 23 October 20000 establishing a framework for Community action in the field of water policy. *Official Journal of the European Communities* ***22.12.2000***: L 327/1 – L 327/73.

Frijns, J., Mulder, M., & Roorda, J. (2008). *Heading for a climate neutral water cycle*. Report KWR Watercycle Research Institute/Stowa Nr. 2008-17: Utrecht, the Netherlands (in Dutch).

Geudens, P. J. J. G., & van Beek, M. J. J. (2007). *Benchmarking in the drinking water sector 2006*. Report VEWIN (Association of Dutch Water Companies) Nr. 2007/79/6292: Rijswijk, the Netherlands (in Dutch).

Guest, J. S., Skerlos, S. J., Barnard, J. L., Beck, M. B., Daigger, G. T., Hilger, H., Jackson, S. J., Karvazy, K., Kelly, L., MacPherson, L., Mihelcic, J. R., Pramanik, A., Raskin, L., Van Loosdrecht, M. C. M., Yeh, D., & Love, N. G. (2009). A new planning and design paradigm to achieve sustainable resource recovery from wastewater. *Environmental Science and Technology, 43*, 6126–6130. doi:10.1021/es9010515.

Haasnoot, M., Middelkoop, H., van Beek, E., & van Deursen, W. P. A. (2009). A method to develop sustainable water management strategies for an uncertain future. *Sustainable Development* Early View Published Online 18 Nov 2009. DOI 10.1002/sd.438.

van der Hoek, J. P. (2010). *Energy in the water cycle*. In J. C. van Dijk, A. R. D. Verliefde, H. van der Laan (Eds.) *Proceedings of the congress 'Energy from water: hype or chance?'* (pp. 37–45). Water Management Academic Press: Delft, the Netherlands (in Dutch).

van der Hoek, J. P., Hofman, J. A. H. M., & Someren, T. C. R. van (2011). Integration and innovation of the urban water cycle: the waternet experience. *Journal of Environmental Science and Engineering, 5*, 533–544.

Keirsted, J., & Leach, M. (2008). Bridging the gaps between theory and practice: a service niche approach to urban sustainability indicators. *Sustainable Development, 16*, 329–340. doi:10.1002/sd.349.

Kiestra, F. (2008). The energy factory. http://www.energiefabriek.com. (in Dutch).

Larsen, T. A., Alder, A. C., Eggen, R. I. L., Maurer, M., & Lienert, J. (2009). Source separation: will we see a paradigm shift in wastewater handling? *Environmental Science and Technology, 43*, 6121–6125. doi:10.1021/es803001r.

Ministry VROM. (2007). *Policy Agreement Water Cycle 2007 ('BWK-2007')*. Policy agreement under auspices of the Ministry of Spatial Planning and the Environment VROM: The Hague, the Netherlands (in Dutch).

Ministry VROM (2009a). *The quality of the drinking water in the Netherlands in 2008*. Report VROM-Inspectorate, publication nr. VROM 7275/RIVM 703179046/2009: The Hague, the Netherlands (in Dutch).

Ministry VROM (2009b). *Connecting water – Long term vision for the Water Cycle*. Report project team "Water Cycle", Ministry VROM: The Hague, the Netherlands (in Dutch).

Moll, R. (2007). *Effects of climate change: research questions for water managers*. Report Stowa Nr. 2007-W08: Utrecht, the Netherlands (in Dutch).

Mons, M. N., van Genderen J., & van Dijk-Looijaard, A. M. (2000). *Inventory of the presence of pharmaceuticals in Dutch water*. Report Kiwa Water Research Nr. 30.3534.011: Nieuwegein, the Netherlands.

North European Benchmarking co-operation (2008). *International Benchmark 2007 - Final report*. Report of the North European Benchmark co-operation supported by IWA International Water Association, www.waterbenchmark.org, November 2008.

Offermans, A., Haasnoot, M., & Valkering, P. (2009). A method to explore social response for sustainable water management strategies under changing conditions. *Sustainable Development* Early View Published Online 18 Nov 2009. DOI 10.1002/sd.439.

Oosterhof, A., Wolthek, N., van der Meer, W., Groenendijk, M., van de Wetering, S., Boukes, H., Raat, K., & Eerhart, J. (2009). Breakthrough in the use of brackish groundwater as alternative source for drinking water supply. *H2O, 42*(14/15), 14–17 (in Dutch).

Otterpohl, R. (2002). Options for alternative types of sewerage and treatment systems directed to improvement of the overall performance. *Water Science and Technology, 45*, 149–158.

Sacher, F., & Stoks, P. G. (2003). *Pharmaceutical Residues in waters in the Netherlands*. Report RIWA (Association of the River Waterworks): Nieuwegein, the Netherlands (in Dutch).

Segrave, A. J. (2006). *Techneau report on trends in the Netherlands*. Report Techneu D 1.1.14 (European Commission, Sixth Framework Programme, Sustainable Development, Global Change and Ecosystems Thematic Priority Item, contract number 018320).

Senge, P., Smith, B., Kruschwitz, N., Laur, J., & Schley, S. (2008). *The necessary revolution – How individuals and organizations Are working together to create a sustainable world*. London/Boston: Nicholas Brealey Publishing.

Someren, T. C. R. van (1992a). Reference framework for Environmental Performance Indicators, Environmental Performance Indicators, European Green Table, March 1992, Oslo, Norway.

Someren, T. C. R. van (1992b). Sustainable Environmental Management. A Market Approach, report for European Green Table, KPMG Environmental Consulting.

Someren, T. C. R. van, van der Kolk, J., ten Have, K., & Calkoen, P. T. (1992c). Bedrijfsmilieuzorgsystemen, Tussenevaluatie 1992, Onderzoeksrapport, KPMG Milieu/IVA, Den Haag/ Tilburg, 1993. (translation of summary available in English: Company Environmental Management Systems Interim Evaluation 1992, Ministry of VROM, nr. 1993/3).

Someren, T. C. R. van (1992d). Innovatie, Emulatie en Strategisch Milieumanagement, Strategisch Milieu Management, MBA-course, Department of Scientific Institute for Environmental Management, University of Amsterdam, Amersfoort, Netherlands, June 1992.

Someren, T. C. R. van (1994). Die Anwendung der EG-Umwelt-Audit-Verordnung in kleinen und mittelständischen Unternehmen. In: Umweltmanagement und Umwelt-Audit in kleinen Unternehmen, IÖW Schriftenreihe 80/94, Berlin 1994.

Someren, T. C. R. van (1995a). Sustainable development and the firm: Organisational innovations and environmental strategy. In: T. C. R. van Someren (Ed.) *Business strategy and the environment* (Vol. 4, No. 1, pp. 23–33), January–March 1995.

Someren, T. C. R. van (1995b). Umwelt-Cost-Controlling. Volkswagen Seminar Umweltcontrolling, November 9, 1995, Braunschweig, Germany.

Someren, T. C. R. van, & van Dam, E. (1997). Financiële instellingen en duurzame ontwikkeling, NIBE, oktober 1997, pp. 14–18

Someren, T. C. R. van (1998a). Strategies for Sustainable Profits, Milieumanagement van Kosten naar Baten, Wetenschappelijk Instituut voor Milieumanagement, Universiteit van Amsterdam, April 7, 1998, Amsterdam, The Netherlands.

Someren, T. C. R. van (1998b). Strategies for sustainable profits, International BMW Seminar, Dietramszell, September 17, 1998.

Someren, T. C. R. van (1998c). Strategies for sustainable profits, International Shell Seminar Managing HSE in the Business, December 7, 1998, Noordwijkerhout, The Netherlands

Someren, T. C. R. van (2005). *Strategische Innovationen. So machen Sie Ihr Unternehmen einzigartig*. Wiesbaden, Germany: Verlag.

Someren, T. C. R. van, & van Someren-Wang, S. (2006). Export van Waterkennis naar China, H2O, No 1, januari 13, 2006.

Someren, T. C. R. van, & Nijhof, A. (2010). *Triple P business development in the Dutch agro-food sector*. Assen: Van Gorcum Publishers.

Unie van waterschappen (2009). *Benchmark Water Boards 2009*. Report Unie van Waterschappen (Association of Water Boards): The Hague, the Netherlands (in Dutch).

Vewin (2008). *Dutch Drinking Water Statistics 2008 – The water cycle from source to tap*. Publication o the Association of Dutch Water Companies (VEWIN) no. 2009/95/6259: Rijswijk, the Netherlands.

Vewin (2010). *Drinking water sector 2.4 percent more efficient*. http://www.vewin.nl/publicaties/waterspiegel/Waterspiegel_2009_2010/Pages/

Zeeman, G., Kujawa, K., De Mes, T., Hernandez, L., De Graaff, M., Abu-Ghunmi, L., Mels, A., Meulman, B., Temmink, H., Buisman, C., Van Lier, J., & Lettinga, G. (2008). Anaerobic treatment as a core technology for energy, nutrients and water recovery from source-separated domestic waste (water). *Water Science & Technology-WST, 57*, 1207–1212. doi:10.216/wst2008.101.

Zwolsman, G., Cirkel, G., Doomen, A., Jalink, M., van den Berg, G., Maas, C., Vreeburg, J., IJpelaar, G., & Mesman, G. (2007). *Risk assessment of the consequences of climate change for the drinking water sector*. Report Kiwa Water Research Nr. BTO 2007.032: Nieuwegein, the Netherlands (in Dutch).

7 Green China Requires New Government

The Government Is Dead, Long Live the Government

Green China implies that Western governments have to play their part as well. Their contribution is in many cases and circumstances bigger than what most Western governments and policy makers are used to. The helplessness of the government in dealing with crises in the USA and EU showed the declining governmental strength. If we want to sustain Western welfare, governments have to be strategically renewed. The government of the USA, the European Commission, and member state governments often have a defensive reaction on the rise of Green China. It is a natural Pavlov reaction. Being unprepared on the speed and scope of changes coming from Green China automatically leads to try to defend your current position and playing the power game by old Western rules. But gradually, Green China leads to a new world order which requires to abandon current government and replace it by a new one capable of being an equal counterpart of Green China government. Western government needs strategic innovations as well. The nature of sustainability requires the involvement of government, and betting on a market innovation approach is not sufficient in the new sustainable growth curve. The USA and the EU have to make tailor-made renewal of the role and contribution of their government.

7.1 European Union or European Onion?

The European Union lacks a growth strategy. Why? Because it is more a European Onion than a European Union. The European Onion means that there are not only several layers but also the layers make one cry when peeling it off. The reason is that the layers do not fit together and do not form a unity. In Fig. 7.1, the European Onion in case of political perspective is presented.

From the perspective of power, the political power European Onion is Europe's "engine." Germany together with France determine course and timing of European palns. Other member states more or less have to follow their proposals and plans.

T.C.R. van Someren and S. van Someren-Wang, *Green China*,
Management for Professionals, DOI 10.1007/978-3-642-28810-4_7,

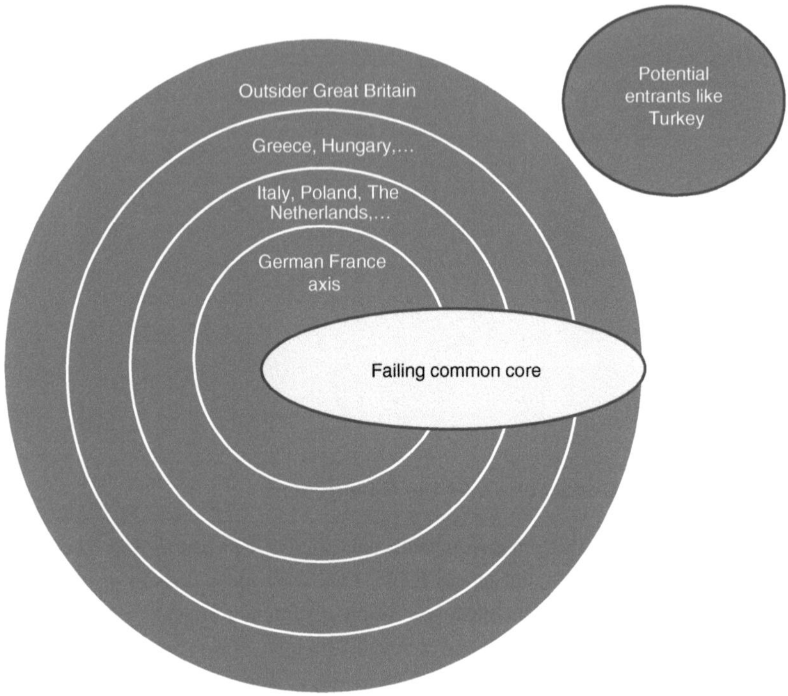

Fig. 7.1 The political European Onion

Since the foundation of the European Coal and Steel Community (ECSC), Europe has achieved a custom union, a monetary union, an economical union, and since December 2011, an attempt to install fresh born fiscal union but no military and no political union. Many argue that when the military and political union would be implemented, the European project is finished. That is a misjudgment. For what reason does an integration project exists? The original reason is to prevent war and to integrate Germany in a larger setting. But 60 years later, a new generation with different experiences and worldview is entering the scenes. And the BRICS emulate the PIGS. Furthermore, with the rise of Green China, a new world is developing which needs new answers. But for an adequate strategic answer on Green China, the integration of Europe is insufficient. The most crucial part in the European integration project is failing. The failing core is a sustainable growth strategy based on shared values, a clear definition of its right to exist, and common aims. Why is it absent? Because we have many other European Onions.

A few illustrations highlight the Onion principle. Since the European summit ended on 9 December 2011, the political layers consist of EU-26 and Great Britain. Another one is the EU-17 + 6. The group of 17 countries organize themselves as the eurozone in which 6 other countries still have the opportunity to join. The unification of Europe is for the current leaders a struggle. The root cause is national interest and failing knowledge about market economy. The political Onion is a system to defend and put forward national interests under the umbrella of integration.

From a governance perspective, also a European Onion can be identified. The European governance consists of decisions taken by the governmental leaders carried out by the European Commission, the individual members, and nation states and regions. Ideally, the decisions taken on a European level should be advantageous for everybody or the whole of the EU. However, national interests often prevent these ideal decisions. Moreover, national interests can be pushed forward because European government is the member states.

With regard to innovation, the European Onion consists of a few member states who currently form the lead group like Germany and Finland followed by a middle group such as the Netherlands and France and a few laggards like Greece and Hungary. Also in green energy policies, the national energy program is regarded to be more important than a pan-European approach and investment plan.

The relevance of these different European Onions is that all current policies are directed to harmonizing European activities and leveling off European differences. For example, in every European Framework Programme (FP), the dissemination of knowledge among member states is an ultimate goal. For building an integrated Europe and bringing together the member states with their different history and development, this might be the right path for making a start. But for getting an answer on Green China, it would be better to accept the European Onion and at the same time, to exploit the differences in entrepreneurship, knowledge, and innovative power. Also, a reversal from funding for consumption to capital for investments is necessary. Keynesian governmental spending must be replaced by pragmatic industrial investments. But when following these inversions in the role of European government, we also need a common overall goal to work on for the coming decades. A suggestion fitting to the main topic of this book would be Sustainable Europe 2050. In the remainder of this chapter, some of the necessary strategic innovations for building a new government and instruments are highlighted.

7.2 Renewal of Government

Current European government and policies of both European Commission and member states are not prepared on a Green China future state of the world. The basic issue is not about the content of policies itself but understanding Green China and its behavior. Only a very few European leaders from government and business understand the behavior of China. Not knowing the current situation and anticipating the big worldwide trends is why most of the European policies do not work. In Fig. 7.2, the sense of urgency is shown.

The megatrends in the world combined with the rise of Green China urge the European Union and its member-stated force to aim for a Sustainable EU in 2050. But this overall goal needs not only strategic innovation of our private companies as demonstrated in earlier chapters. It also requires strategic innovation in government, policies, and public organizations as well. Why?

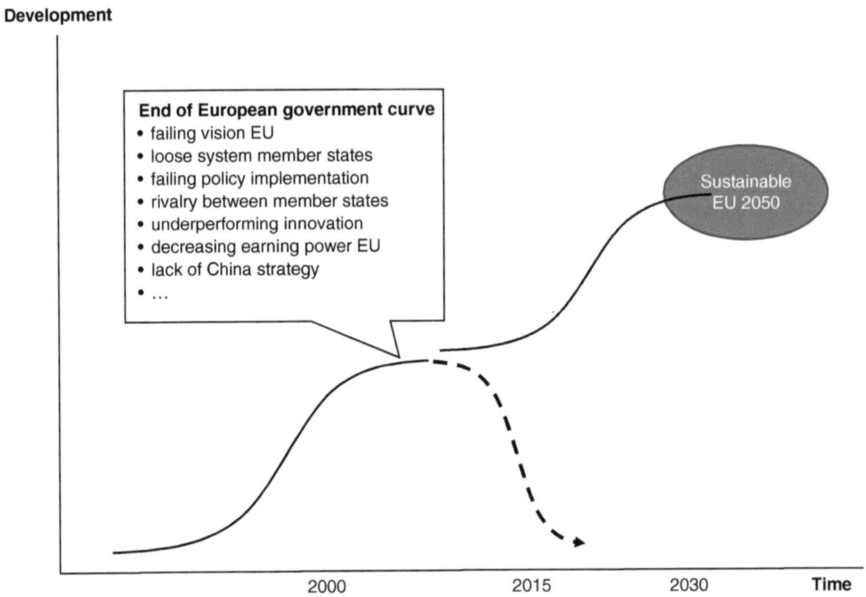

Fig. 7.2 End of current European government curve. (Source: Based upon van Someren, 2005, Strategische Innovationen, Gabler Verlag)

The shift of political and economic power accompanied by new market rules, new players with different business practices, and new governmental roles requires a fundamental restructuring of European government and policies. We all received these signals for over a decade, but we were not able or not willing to get the big picture. We enjoyed our welfare states, and we were bothered with minor problems and improving the existing system. We were so proud on our Europe project that we got a Eurocentric worldview. Due to the Eurocentric worldview and inward-looking attitude, we missed and neglected the strong and weak signals announcing major changes. We give a few examples. There is no European vision on the rise of BRIC countries. European vision was limited to expanding the borders of Europe, and setting up rules and governmental structures like Europe was an island. The loose system of member states was covered by the excuse that Europe needed its diversity for the sake of diversity alone. No strategic plans were made to make the loose system work outside of Europe. Or how to exploit it. Plans were only made for internal transfer of money, goods, and labor. Or no plans were made for how the EU would fit in a changing world order. The underperforming innovation rate and output has been known for a long time. Average R&D spending, but more importantly innovation output, stabilized for over many years. We must conclude that most European and national innovation policies did not work. There was no uproar after the disaster of the Lisbon strategy aiming at Europe being the most competitive region in the world. Why was there no heated discussion? Why were there no street protests? Why did industry not protest loudly? Because it was thought that it did not matter or we had time to fix it. Policies had little effect such as innovation policy. Policies tried to improve the existing system such as internal tax treaties, labor conditions, and internal competition. Politicians were being paid, civil servants got their salaries and pensions, bank managers got their bonuses, and the earth still turned around the sun. We lived in luxury, and there was no sense of urgency. Now, there is. Green China forces us to act very quickly and boldly. We think when the euro crisis has been solved, Europe is saved. Chancellor Merkel said that if we were not able to save the euro, Europe will be lost. That will not be the case.

Europe will only be saved if we are able to create governmental roles, policies, institutions, and businesses to create a second curve realizing Sustainable Europe 2050. The only way out is to create a society able to generate the value conversion engine needed to achieve Sustainable Europe 2050. Only businesses and governmental policies which lead to new sustainable earning capacity within the context of Green China will save the EU. The financial system and the euro are additional instruments creating favorable conditions. The next section focuses on the main elements of governmental instruments needed for Sustainable Europe 2050.

7.3 The Wrong Power of the Superpower

The USA, together with the former USSR, was the superpower of the last century. After the cold war and the disintegration of the Soviet Union, the USA was the only one on the Olympus of political, economical, and military power. It was indeed lonely at the top, and in these situations, despite the perfect view of the world, it depends how the superpower is used. At the top of the growth curve, some incidents indicated a possible decline like the Gulf war, 9/11, war on terrorism, and financial crises. Meanwhile, China was moving toward the top, and simultaneously, the long-denied issue of sustainability gained power in the international community. In the coming decades, the current superpower USA has two major challenges to tackle: China and greening of the American industry and society. It depends on the strategic innovations on policy and corporate strategies whether it will turn into another American dream or a nightmare.

7.3.1 American Green Dreams or a Sustainable Nightmare

In the recent past, both US government and private industry have paid relatively low attention to strategically important sustainability issues. Former Vice-President Al Gore was able in old Hollywood tradition to make a movie about the threats of climate change. But the USA is missing the opportunities. Both the US government and large parts of private business are lagging behind the EU and some of its member states. Especially the opportunities of sustainability are not recognized. In the past years, the US government and its Environmental Protection Agency (EPA) just did the expected law making and regulations on all relevant fields like air, water, and soil. Moreover, the states have a relatively large power to make their

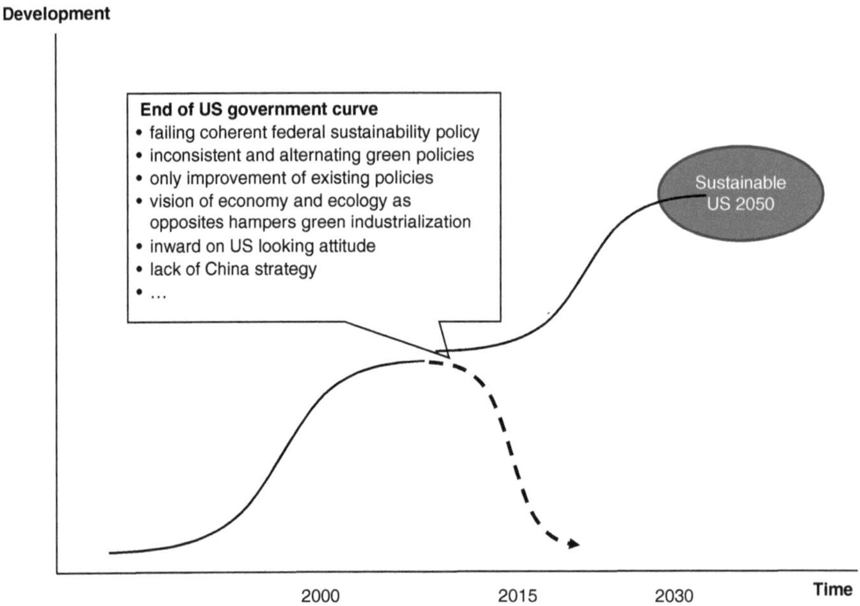

Fig. 7.3 Strategic innovation of US sustainability curve needed. (Source: based upon van Someren, 2005, Strategische Innovationen, Gabler Verlag)

own regulations, and some take the chance to do some more. For example, California is pressing forward by introducing environmental laws forcing industry to act. But in general, the American laws and regulations are characterized by improvement and alongside the current growth curve of existing industries. When coordinated and continued consistently on the long term, intelligent laws and regulations can trigger the development of new industries in the field of sustainable development. Germany has proved that. But these kind of radical innovation in the field of sustainability is largely absent in the USA. The result of this non-innovative attitude is shown in Fig. 7.3.

Why is the USA with their huge reputation of being entrepreneurial and innovative so conservative with regard to sustainability? There are several reasons. The most important is strategic one. In the past decades, the dominant view among both political and business leaders was and still is that economy and ecology are opposites. Being green was something leftist and a non-heroic topic far away from hard-core business like high-tech or investment banking. Only in recent years, the eyes are wide opened at least in the investment world. But there is a far way to go. The failing coherent and long-term legislation as a stimulus for renewal as in Germany is another reason for lagging behind. The inward-looking attitude of the USA is a consequence of sitting lonely on the top of the mountain. Being occupied by oil and terror wars, the geo-economic-political rise of China was disregarded by US government in the recent years. The US government needs a megastrategic innovation program to catch up with worldwide developments.

President Obama took some action and let us see whether it has the substance needed to become a major player in the future sustainable world.

The USA wants to reclaim its world dominant position and to sustain superpower status. This is the core of President Obama's "Strategy for American Innovation" policy issued in 2009 and updated in 2011 (White House 2011). The aim is to lay the foundation for the innovation economy of the future. With regard to the topic of sustainability, the clean energy revolution and the electrical vehicle program including battery and biofuel technology have priority. The biggest investment in the 2009 program is allocated to renewable energy and efficiency sector with a budget of over $30 billion. The second place is for health IT with a budget of nearly $20 billion. Advanced vehicles and biofuels get approximately less than $7 billion. In the updated policy of 2011, it is mentioned that "To win the future, we must out-innovate, out-educate, and out-build the rest of the world." (White House update, 2011). Or in one soundbite: emulate the rest of the world. With regard to energy, President Obama has proposed "...a Clean Energy Standard that will help us reach a goal of delivering 80 % of the nation's electricity from clean sources by 2035." (White House update, 2011). The aim for 2015 is to have one million advanced technology vehicles on the road. Moreover, the Administration of Obama allocates budget to set up, firstly, the Advanced Research Projects Agency-Energy (ARPA-E) and secondly, to create three new Energy Innovations Hubs to solve challenges in critical areas. Other complementary measures for the renewable energy sector are a reauthorization of the clean energy manufacturing tax credit and budgets for research, development, and deployment in the advanced technology vehicles program.

These measures are part of a bigger innovation program which will not be discussed here. But the sustainable policy being part of the larger innovation policy is not very innovative or far reaching. Again, a missed opportunity. The USA falls back on the old success model of the market-based innovation approach, whereas another approach is needed to make sustainability a success. In Table 7.1, some of the differences are summed.

In the past, in many industries, but not all, the USA outperforms many other countries or regions like the EU. The USA, both public and private, spends a larger portion of their GDP on R&D than others. Also, in the field of nontechnical innovations, US firms have often been front-runners like in fast food, containerization, and supermarkets. But there are also signs that strategic challenges are being missed like water management, hybrid vehicle technology, or battery technology. Also, the USA has to make choices where to excel. But the historical successes are based on innovative entrepreneurs and market-based innovation. The latter is still the main element of Obama's "Strategy for American Innovation." This approach fits to consumer products like the gadgets of Apple, but not to sustainability challenges. The reason is that the IT sector is full of small SMEs trying to invent a new market, and short-term success is key. If they fail, it is no problem; others will take over. The successful ones can cluster like in Silicon Valley. There, creative entrepreneurism flourishes as nowhere else without the interference of government.

Table 7.1 Difference between traditional American innovation approach and sustainable innovation

American green tech silicon valley approach	Sustainable innovation
Focus on radical new technology	Focus on strategic innovations
Pool of SMEs focused on product innovation	Strategic innovations in combination with complementary infrastructure
Low-regulated market	High-regulated market
Apart from environmental laws, sustainable laws, and regulation largely missing	Laws and regulations as change agents
Quick short-term buck	Slow long-term buck
Low cost	Large investment market
Reliance on private market and market-based innovation	Symbiosis of private and public sector
High founding, high turnover, and failure rate of firms due to low-cost low-entry barriers and large supply of entrepreneurs	Slow founding, low turnover and low failure rate of firms due to high-cost high-entry barriers and low supply of firms

This is in a nutshell, American capitalism. But why does American capitalism not work in the start-up of the sustainable industries?

This Silicon Valley model has also been applied in the field of sustainability. Silicon Valley venture capitalists have invested over 30 billion in sustainable projects, but all the money has been vaporized. In one occasion, Solyndra, the US government took the very unconventional role of picking the winner which ended in a disaster. Solyndra is a showcase of failing governmental sustainable policy and lack of a China strategy. Solyndra seemed to be a start-up winner in the solar panel industry and received governmental capital support, a $535 million loan guarantee against favorable conditions and low-interest rates of 1 % to 2 %. Still, they were not able to compete with Chinese products and went bankrupt in autumn 2011. Market conditions were very volatile; in 2011, prices plummeted with 70 %. And the Chinese government supported the solar industry with a total of $30 billion. In these circumstances, market innovation goes bankrupt. But there is more. The failing sustainable laws and regulations together with the absence of a long-term strategy for creating the conditions of new green industries is the exact opposite of nongovernmental intervention and short-term behavior in the Silicon Valley approach. It is also the quick buck against the slow long-term buck. Greening industries not only requires new firms but large investments in countrywide infrastructure as well. The latter is a task of government to at least coordinate or even invest.

Despite European and German laws and regulations and the buildup of a solar industry in Germany, the Chinese solar industry profits from German tax money. The same happens in the USA. Therefore, strategic policies for sustainability are not enough. An American–China policy and strategy is a parallel necessity.

In Table 7.2, a few highlights of the current situation in the US market demonstrate the differences in the approach and the resulting market outcome.

The core question is who will earn the green buck? Contrary to Chinese laws and regulations, the US laws and regulations are linked to the current growth curve

Table 7.2 Who will earn the Green buck

	USA	China
Regulation	Current growth cycle laws and regulation	New growth cycle creating laws and regulation
Market	New technology-based small markets	Large-scale home market and exporting overcapacity of proven technology
People	Melting pot of best talent	Sea turtles and domestic small dragons
R&D	Fundamental R&D	Proven technology improvements
Business innovation	New technology product innovation	Low-cost manufacturing innovations and product adaptations
Funding	Market and venture capitalist	Government supported
Time horizon	Short term	Short term and long term

instead of the new future growth curve. Chinese laws and regulations are not made to make industrial life difficult but to simultaneously create a market and protect the environment. American laws focus on protecting the environment and health of the people. The Chinese exports of, for example, low priced solar panels are based on a bunch of mutual reinforcing factors: high-quality proven technology, economies of scale (overcapapcity) driven low cost, economies of scope-driven other appliances of solar technology, economies of time driven 24/7 factory times and high flexibility, listed and access to American stock exchange, and government supported favorable bank loans. The big asset of the USA is its entrepreneurial creative people which are mostly a real melting pot of talent. The Chinese fall back more on their sea turtles, the expats who not always return home, and the domestic small dragons with their own businesses. The overcapacity is a logical consequence.

7.3.2 American Yellow Dreams or a Red Nightmare

The famous painting of the American artist Barnett Newman "Who is afraid of red, yellow and blue" reflects the relation between both the USA and China and the USA and the EU. In the painting of Newman, the dominant color is red, but also the red surface overwhelms in size the blue part. In geo-economical-political terms: the USA is not afraid of blue EU. With regard to China, a different picture emerges. Depending on the ruling party in the USA, there is a dream or a nightmare. When the blue Democrats are in power, the dominant view is that China is more a threat than an opportunity. Then the USA is really afraid of the Yellow Danger. The opposite is the case when the Republicans are in power. The Republicans always had a good pragmatic relationship based in business opportunities and power play with Red China. Nobody was afraid of red. These notions are also shared by the Chinese. And it is interesting to observe how China is preparing itself on competing with the USA.

China admires the USA for its economic power and innovation ability. China respects the power of the USA and always had the desire to economically cooperate with the USA. Therefore, it is not a surprise that in most Chinese Foreign Affairs

Offices (FAO) of Chinese central and local government, the department focusing on the USA is the biggest one. Bigger than the ones occupied with the EU. But the cooperation did not take off. One of the main reasons is that a comprehensive economic development US policy on China is largely absent and former ad hoc policies have been unstable with alternating views and standpoints. In the future world, this has to be changed. President Obama announced to withdraw from a two-country war to focus on naval and air power in especially the Pacific region. This is nothing else than reacting on shifting global powers and a typical American way of reformulating the country's interests. With this strategic move, the focus is still on the diplomatic-political-military level. There is still no economic China policy of the same order. Who said: "It's the economy stupid!"? Exactly, former President Clinton and future Administrations should strategically pay attention to the Chinese economy and its fire power in the green industry complex.

From a Chinese governmental point of view, the combination of centralized and decentralized government together with private industry, the USA has an enormous power. But the power is wrongly used and applied. The American fixation on military power and policies of economic confrontation make the USA blind for future challenges and new world order rules. Private industry uses economic power to dominate the old industries. But for the new industries like renewables and eco-mobility, a new long-term green deal with government is necessary.

7.4 Will Green China Eat, Beat, or Treat the West?

Of course the US administration, European Commission, national European governments, political leaders, and policy makers acknowledge the rise of China. They all understand China has the opportunity to become the biggest economic power within the next generation. With regard to the economic angle, all kinds of statistics, studies about trade volumes, and deficits are being made and discussed. The response is to put issues on the agenda like intensifying trade talks, discuss the imbalance in trade including the role of the RMB, IP rights, intensifying relations, and of course human rights. What is wrong about it? The topics might be right, but Europe encounters China with Western eyes. This is natural; as European, we think like Westerners because we are born here, raised here, and had our education and working experience here. That makes it difficult to see China with Chinese eyes as Chapter 4 has showed. And that is exactly necessary. Do we have to become Chinese? No, of course not, it is even impossible. But the West should follow Sun Tzu advice. Sun Tzu said that "if you know your enemies and know yourself, you can win a hundred battles without a single loss. If you only know yourself, but not your opponent, you may win or may lose. If you know neither yourself nor your enemy, you will always endanger yourself." Europe and the USA need Chinese eyes not to become Chinese but to understand the underlying strategic behavior and attitude of China. It is only an instrument to defend Europe's own interests. But we do not only need Sun Tzu. Also, Aristotle already said that the easiest thing in life is to give advice and the most difficult one to know yourself. He is still right. Europe knows exactly what our countries have to do, but we are not able to define or describe our European values. This attitude comes back in Europe's current behavior toward Green China. The USA has to abandon its inner looking policies. The USA is still alone on its own top of the mountain, but the surroundings have changed. Volcanic eruptions have created new mountains around the USA, and the eruptions are not over yet.

On the short term, China needs both the EU and USA because of the export, available technology, and foreign investment opportunities. Besides exports, China will make a switch from exports to domestic consumption, but that will take time. Therefore, the Green Dragons of China will not eat, beat, or swallow Europe or the USA in one bite on the short term. The EU can make use of this window of time to define its role in the world and to find a new sound basis for its legitimacy. Based on this new footing, a China strategy has to be formulated. If the EU is not able to define its right of existence and to formulate its legitimacy, we will defeat ourselves if we are not defeated by upcoming countries like China. Like the frog in the pan who asks if the water is already getting hot and is too late to get out when it is really cooking. The USA has to change its focus from unilateral power to coping with a new multilateral world.

Private business and public sectors have the identical challenge of making a jump to a new curve of prosperity, growth, and development. Green China forces a strategic innovation of the European Union and the USA. The governmental jump can be described along the four core areas around the business models. In the Western context, creating new business models maintains the core task of private businesses. However, with the rise of Green China and the aim of building a sustainable society, the four areas of leadership, institutional innovation, business development, and governance remain relevant. But from a governmental perspective, they get a different content. In Fig. 7.4, these four elements are described from a governmental and policy perspective.

The governmental strategic innovations should create favorable conditions and support the initiatives of the private sector toward a sustainable Europe 2050 by means of green European business models. The strategic innovation journey is a possible way to create new governmental growth curve. The four elements of governmental strategic innovations can be a start of this journey. But in several instances, the journey for the USA will be different than for the EU.

7.5 Leadership

Lack of leadership in Europe has cursed us enough troubles. Europe's crisis is a crisis of its right to exist and legitimacy; not of debts, banks, or euro troubles. The road of the EU is one of an economic union toward a kind of political union. Whatever this path to the future will bring, already the existing economic union and its borders with Russia and the Islamic countries like Turkey requires a global definition of its role in the world. On top of that, the economic relation with and the growing importance of China requires a China strategy. The way the EU defines its

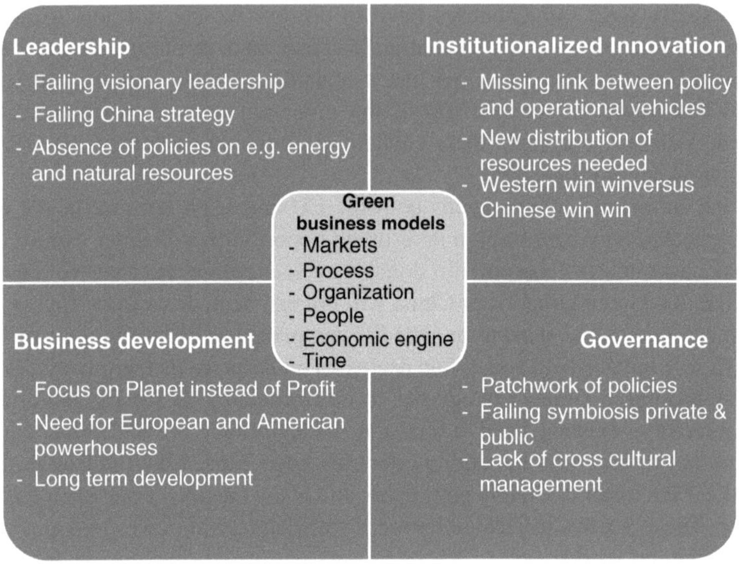

Fig. 7.4 Governmental strategic innovations

right to exist and wants to exploit as an economic power determines whether Green China will eat the EU or treat it as a respected counterpart.

The rise of Green China in the past decades is not only a power shift but also a shift in ways of doing business. The immediate implication is that not only changes in politically dominated power forums like IMF, World Bank, G20, and WTO take place but also in industries. The European answer on both fronts is inadequate. 2011 marks the 10th anniversary of the membership of China to the WTO. China has profited from global exporting and the EU from cheap Chinese products, low manufacturing costs, and even the booming marketed of capital goods like cars and technology. But it may not be long before it turns into one-way traffic if the EU will not pay attention to a good China strategy and a coherent industrial and innovation policy.

The growing Green China successes should be a wake-up call for European policy makers. What is the common reaction of policy makers? Our Pavlov reaction is defending our world order by putting issues on the agenda which does not recognize the Chinese policies and their impact on the economy but represent a defense wall against China. China knows as no one other that building great walls is useless for foreign threats. For example, the 6th EU-China Business Summit 2010 mentions the common aim of "innovation as a win-win solution for steady growth," and that the Doha round should lead to an accession on governmental procurement and lastly, that counterfeiting and piracy is a global problem and China and the EU should cooperate bilaterally and multilaterally (EU-China Business Summit 2010).

The same holds for American–Chinese trade talks. The American Secretary Tim Geithner remarked before the opening of the U.S.-China Strategic and Economic

Dialogue in 2011: "Our great strengths as a country have been in our openness to ideas and talent, our capacity to innovate, our excellence in higher education, a willingness to invest public resources strategically in scientific research and discovery, and the political will to confront challenges with wisdom and force" (U.S.-China Strategic and Economic Dialogue 2011a).

But reading between the lines brings a totally different picture. How? We take both examples of the European Commission and the USA. These dialogues show the different thinking and assumptions on each other's behavior and strategic thinking.

7.5.1 Europe–China Trade Dialogue

The same line of thinking as in the China-European Summit meetings can be found at the European Commissioner for Trade who puts the following issues on the agenda in trade talks with China (De Gucht 2011):

- *"An open business climate:* there is a general feeling in Europe that economic openness in China is not improving. The recent annual report from the EU Chamber of Commerce in China seems to confirm that the business climate in China gets worse, reflecting a severe *imbalance in market access* and significant behind-the-border issues.
- *Investments* in China, where key sectors remain closed or subject to ownership restrictions. This blocks significant economic gains on both sides from being realized. It erodes the platform for explorative talks on a possible investment agreement. It also plays into the hands of those in Europe who see Chinese investments as a threat. So far, we have kept this debate at bay by maintaining that Europe's open investment regime remains our strongest argument for others to grant us similar access.
- *Protection of Intellectual Property*: Significant infringements undermine not only European investments and technology transfer to China, but just as much China's ambitions to become an innovative economy.
- *Procurement*: We keep hearing complaints from European businesses that the Chinese procurement market is closed and lacks transparency and regularity. This has also spurred a debate in Europe to strengthen '*reciprocity*' in our external economic relations.
- *Raw Materials*. Hopefully, the ongoing WTO case will contribute to undistorted access to raw materials in China. As a major importer of raw materials itself, China also has an interest in maintaining open and non-discriminatory global access to raw materials, including amongst others iron ore and cotton.
- *Subsidies and State Owned Enterprises*. These are issues which are rapidly rising in importance, and there are increasing calls in Europe to address them to restore more level playing field to our economic operators. To deal with this situation we need to engage in dialogue on these issues. More transparency on them would also be very helpful."

These topics are short-term issues which should enable a new long-term relation with China. These short-term agenda points itself are not wrong or wrongly timed. The issue is how do you deal with them in order to serve your own interests without keeping an eye on the interest of the other. These issues are presented and interpreted from a Western perspective and wrong assumptions about Chinese behavior. You as a reader of previous chapters should meanwhile be able to identify that our perception of Chinese behavior is going in the wrong direction. As a consequence, we are not able to defend our interests properly. Where does it go wrong? The interpretation of these points should be as follows:

- *An Open Business Climate*: Opening the Chinese market for foreign companies under Chinese conditions is official policy to build up their Green Dragons. For example, China does not only want to be the factory of the world; pollution intensive industry is discouraged almost everywhere. On the other hand, high-tech and sustainable industries are enjoying the most favorite treatment in most of the economic zones. Hence, openness is not a mantra but open and close are used which one is most favorable to Chinese development.
- *Investments*: There are two kinds of investments eager to flow to Europe. The governmental investment is most of time strategic and aiming at secure important raw material, market, or technologies. The private ones are most of the time looking for financial or personal chances. As long as the Europeans need the large cash reserve in China, we need a better way of attracting the right investment. Rather than only look at the positive part like jobs or only as threat like taking away technology.
- *Protection of Intellectual Property*: China wants to be an innovative country itself. As long as they do not have technology themselves, eyes are shut at IP rights. Once they make noises about protecting a certain IP right, it is mostly sure that they learn from the different sources and developed their own unique one combining all the existing ones. It is Sun Tzu's strategy of learning in the West and surpassing in the East which drives Chinese policy makers. As soon as China has reached the level of being an innovative China, IP rights will be very high on the agenda. Probably much higher than the Europeans want.
- *Procurement*: China will change procurement rules and make them more transparent but simultaneously will favor Chinese organizations. If we want to have a chance, we have to have something they do not have. The EU cannot ignore, remove, and replace 5,000-year-old guanxi.
- *Raw Materials*: Here, the Western idea is based on free trade of raw materials. This is a naive and unrealistic interpretation. China will try, as they already do, to capture foreign raw material sources. They use several instruments like buying mines and by means of vertical integration of private- or state-owned organizations. Chinese raw materials like rare earth will be protected.
- *Subsidies and State-Owned Enterprises*: There will be less subsidy and less state-owned enterprises in nonstrategic sectors in the long term. But the Chinese will not give up their plan economy, especially after seeing what has happened in Western countries when the market has gone too far. There will be no complete transparency or openness in the key economic vehicles of central government.

Does the European Commission know everything about the corporate strategies of our key economic vehicles in this case the private companies? No, of course not. Why ask it from others?

7.5.1.1 US–China Trade Talks

The US–China talks are institutionalized in the U.S.-China Strategic and Economic Dialogue and started in 2006. Apart from the hot topic of financial crisis, these dialogues were mainly triggered by four issues: the trade deficit, the value of the RMB, access to Chinese markets, and counterfeiting.

To put it bluntly, when summing up the issues of the U.S.-China Strategic and Economic Dialogue as of 2006, the American way of dialogue is to finger point at the Chinese. Despite political and diplomatic phrasing and agreed areas of cooperation, the American approach is forcefully asking the Chinese to change their policies.

Appreciation of the RMB: With regard to the Chinese currency, the USA is complaining about the artificially low value of the RMB compared to the dollar. The low RMB fires up the exports to the USA and wipes away American manufacturing and American jobs. For years, the USA asks the Chinese to appreciate the RMB.

Counterfeiting: The copy–paste of knowledge and business formats makes competition with Chinese organizations difficult. The protection of IP rights and also business formats should be taken care by Chinese government. In the dialogue of October 2011, it was agreed that "China pledged to improve its high-level, long-term intellectual property rights (IPR) protection and enforcement mechanism, building on the current Special Campaign Against IPR Infringement and Fake and Shoddy Products" (U.S.-China Strategic and Economic Dialogue 2011b).

Trade Deficit: Since 1985, the trade deficit of the USA with China has increased from $6 million in 1985 to $272,321 million in 2011 (US Consensus Bureau 2012). The access to Chinese markets by American firms is hampered by protectionistic rules of Chinese government like the forced cooperation with Chinese firms. In trade terms, like the EU Commission, a level playing field is absent according to the American view. In a testimony of Treasury Secretary Timothy Geithner before the American Senate Finance Committee on 6 October 2010, the first American successes were mentioned: "China has also committed to revising its investment catalog and to encourage and expand areas that are open to foreign investment, including certain services, high-technology goods, high-end manufacturing, and energy saving products. Finally, China also agreed to facilitate foreign investment by narrowing the scope of its investment review and approval processes" (Geithner 2010). This was followed by the agreement in 2011 on the issue of "Competitive Neutrality for Private Firms and State-owned Enterprises": "China and the United States discussed the principle of equivalent treatment for state-owned, controlled, or invested enterprises (SOEs), private enterprises, and foreign enterprises with respect to access to credit, tax treatment, regulatory applicability, and access to factors of production. The two countries also discussed the desirability of ensuring that SOEs seek a commercial rate of return and steadily increase their dividend payout" (U.S.-China Strategic and Economic Dialogue 2011b).

Balanced Growth and Investment: "For many years, China's economic strategy was dependent on rapid export growth. Today, China is committed to transforming its economy into one where growth is generated by home-grown demand and Chinese household consumption. China has also committed in the G-20 to reducing trade and current account imbalances. A consumption-driven Chinese economy will create more opportunities for U.S. firms to export to China, and is a critical part of ensuring strong, sustainable, and balanced global growth" U.S.-China Strategic and Economic Dialogue 2011b).

The rather direct confrontation with the Chinese counterparts fits into the American way of fighting phrased as you are for us or against us. But the six-million-dollar question is whether the cowboy can win from the dragon? A way to find out is to look at the bottom line result of the dialogues and interpret them from the perspective of Green China. Again, just as in the European situation, where does it go wrong?

Appreciation of the RMB: China will appreciate the RMB, but the timing will depend on the moment China has a strong position. According to Zhong Wei from the Beijing Normal University, "the benefits to the United States of RMB appreciation are nearly zero. . .Cheap Chinese goods have been a subsidy for the poor in the U.S., and now the U.S. government wants to eliminate such subsidy while it's having difficulty creating jobs" (Zhong Wei 2011). Already, now the labor costs in China are rising and manufacturing in the USA is for some industries becoming a real option. Therefore, the focus should be less on the currency debate and more on building up a competitive innovative industrial base. Germany shows that high-innovative quality products sell and in which currency is of minor importance.

Trade Deficit, Balanced Growth, and Investment: In the dialogues at the annual U.S.-China Joint Commission on Commerce and Trade Chinese Vice-Premier Wang Qishan said: "An unbalanced recovery would be better than a balanced recession." As we have seen before, China is investing heavily in green industries. The discussions about counterfeiting are not unjustified, but they are linked to the old growth curves. The new industrial growth curves are linked to green industries and services. The assumption and requirement of the USA that Chinese should switch from export-based economy to a consumption-based economy is short-sighted. The relatively minor issue is the necessary investments of the USA in their infrastructure such as roads and high-speed railways and health care. The European Union has granted aid and provided guarantees of € 4,6 trillion to the financial sector (Barroso 2011). The real major issue is the buildup of new strategic sectors. In the period between 2011 and 2016, China invests 10 trillion RMB ($1,7 trillion) in new strategic sectors. China does not worry about imbalances; it plans for future Chinese sustainable growth and domination of rising (green) industries. It is again the failing American response to the world's challenges of this century which is the real imbalance.

Looking with Chinese eyes helps to understand the Chinese position, only then is formulating a winning China strategy of the European Union and USA possible.

Both the USA and EU need to have a nationwide and pan-European strategy to cope with the rising power of sustainable China.

We need leaders with vision for the future, not only like financial fire fighters who move from one euro summit crisis meeting to the other but who are able to tell the European people what happens in the world, what is the relevance of Green China, and its relevance for Europe. Who will stimulate the needed strategic innovation of private and public organizations and lead us to overcome the challenges to contribute what we can toward a sustainable Europe 2050 and a sustainable US 2050.

7.6 Institutionalized Innovation

7.6.1 Missing Link Between Policy and Operational Vehicles

In both the European Union and USA, there is a missing link between policies and bottom line results. It is not only about an underdeveloped triple helix structure or other difficult organizational structures. In the USA, the venture capitalists play a bigger role than in the EU. But with regard to greening industries, it is about commercial organizations who should be immediately responsible for carrying out or picking up policy aims. Here, both economic powers have their difficulties. Even Germany. For example, if a (local) government stimulates solar panel energy sector, it should foster the installation of panels in its region. A classical way of creating a home market. On paper, this implementation is foreseen, but the numerous rules and regulations often hinder a fast implementation. Especially for SMEs, these time-consuming regulations are deadly and not profitable. Good policies are only effective when good implementation policies support the aims. In Europe, this link is often absent and nobody cares about the missing link between technology policies and realization as the following case shows.

The German strength is the relatively strong innovative capacity of both their MNEs and SMEs ("Mittelstand") combined with their focus on technology. The technology focus fits very well to the interest of the Chinese. Almost a common language exists because many Chinese governmental and business leaders have an educational background in technology. But as we have seen in the previous chapter, technology can lead to a technology tunnel focus leaving aside commercial interests and cost-benefit analysis. What to think about the state-of-the-art Maglev technology ("Magnetschwebebahn") developed in cooperation by Siemens and Thyssen Krupp. This new fast magnet propulsion train technology has been developed by these two companies, some affiliated organizations and institutions, and government. A true triple helix. But it was impossible for the Germans to create a home market. Some potential launching trajectories like the connection between Berlin and Hamburg and the center city of Munich with Munich airport were rejected because of cost-benefit discussions and objections by the public along the planned track. Instead Siemens and Thyssen Krupp went to China who of course welcomed them. The end result is a well-known track between Shanghai Pudong and Shanghai International Airport. Moreover, the Chinese now possess all the relevant patents, and prototype copies made by different Chinese manufacturers already exist. The same procedure was followed by the fast train technology. All main manufacturers were invited to cooperate with the Chinese. The Japanese with their bullet train, the French with their TGV, and the Germans with their ICE. The Chinese cooperated with them on different fields. But the Chinese divided their cooperation aims and projects in such a way that the full technology blueprint could be put together. In the EU, national governments promote smart solutions for everything but have no idea how smart should be concretized. The Chinese are smart with copying and giving the feeling that Europeans can do business with good prospects. In most instances, the good feeling disappears and the business opportunities as well. Therefore, our current institutionalized innovation system does generate inventions, but at least in China, profits of western inventions in the long run will evaporate.

7.6.2 Resource Allocation

The past decades, European Union has invested a lot in defending interest of specific industries and regions. Over decades, the agricultural sector has received a disproportional part of available European budgets. Also, regions and new members received large sums of money for building up infrastructure and strengthening the economy. Investments in innovation were largely focused on R&D and knowledge creation and much less on developing markets and industries. In every round about budgets, the reallocation of money proves to be very difficult, and lobbyists are capable to avoid budget transfers or cuts from old growth curve industries and issues. An improvement curve is being created.

Green China requires a reallocation of budgets toward building up industries resulting in world class players. For example, structural funds for any kind of project in Eastern Europe should be allocated to a cluster of European solar energy industry.

In the US market, innovations were the driving force for renewing industries. Exceptions are the military and space sectors. Silicon Valley is the big success story of the USA. Both the EU and China tried to copy it. With regard to green industries, in the USA, some new clusters are emerging like Energy Regional Innovation Cluster, or E-RIC, which will develop technologies for energy efficient buildings. But the question is whether it will be enough in relation to the Chinese investments.

7.6.3 Western Versus Chinese Win–Win

From a Western perspective, the economic rise of China is reflected in a necessary integration in the world economy. In fact, it is the integration in the current Western system based on our norms and values, institutions, laws, and regulations. The idea is that when China adopts to our system and shares the necessary responsibilities, a win–win is possible. Some of these responsibilities to be shared are the exchange rate of the RMB, intellectual property rights, CO_2 emissions and other climate change issues, and human rights. Another aspect fitting to the win–win interpretation is the call for openness. The idea of open innovation or open societies is a myopic view and only applicable in specific circumstances of being behind, sharing risks, and costs. This view neglects the Chinese view on the course of the world. In Chinese eyes, a new world order will emerge in which not only the West will dominate the norms and values, institutions, economic, and financial organizations. But China will not act like a superpower until it is powerful enough and when it is necessary.

Green China may dominate the future world if we are not prepared. If China takes the lead and responsibilities for a sustainable world and their win–win is not based on a benefit for all but first for China, in Chinese eyes, there will be no one world but a Green China as the center of the world. The strive for power and hierarchical relations will make Green China the big boss. This is the ideal harmonious world in the Chinese way. The Chinese interpretation of the win–win can be seen in the way how Western technology and IP rights are being adopted, imitated, and improved. China smartly applies another Sun Tzu warfare practice: Use the strength of the enemy to defeat your enemy. This attitude will not change because China is becoming more important in the world economy and will share responsibilities. The result? According to the European Commissioner of Trade de Gucht: "Over the last decade, European companies have contributed hugely to China's economic development with investments, jobs, and being the largest contributor to China of transfer of technology."[1] His American colleague Geithner said, "We are willing to continue to make progress on these and other issues that matter to China, but our ability to do so depends on how much progress we see from China on the issues that matter most to us. Of course, our ability to take advantage of the economic opportunities and to meet the challenges in this relationship depends on our ability in the United States to deal with our economic challenges here at home" (Geithner 2010).

These sentences can be interpreted in two ways. First, the West can be proud on its long relation with China. Second, who is benefiting from these transfers and relations on the short and long term?

China intensifies its direct investments in the world. Instead of buying state bonds, foreign direct investments strengthen the Green Dragons abroad, and the EU and their member states are, among other regions in the world, one of the key targets. Green China not only buys infrastructure like the port of Piraeus in Greece but also is involved in building European infrastructure like roads. The next step on this path is to further increase the acquisition of European green tech companies and getting involved in research organizations. At the same time, China will increase its support of organizations like IMF. But at a price. The price will be to speed up the

[1] K. de Gucht, 2011, Looking beyond the crisis: Making the EU-China trade relations work, EU-China High Level Political Forum 2011, Brussels, 8 November 2011.

recognition of China as a market economy status within the WTO. This will enable China to export against more favorable conditions. Again a triple Chinese win–win is in the make. Its influence in world institutions will grow, its export base will be strengthened, its access to knowledge and markets will be increased, and rivals will be attacked.

7.7 Business Development

With regard to climate issues, the aim of the European Commission is focused on ecological issues such as CO_2 emission and noise levels. The USA is focusing on the issue of energy independency. That sounds logical. It is not. Why? Because environment sustainability is a business and a profit or better Triple P value generator (van Someren 1995, 1998a, b). In this respect, one should look at the climate conference in Copenhagen (2009) and its follow-up in Durban (2011). The Europeans were thinking the conference was about climate and saving our world. In Copenhagen, during the closing ceremony several participants became emotional when announcing that an agreement had been reached. The emotion was based on the idea that some European leaders believed they were contributing to saving the world. The last-minute agreement was because of negotiations with the Chinese. For China, it is not only about saving the world. For the Chinese, the core issue at the Copenhagen conference was about securing the existing manufacturing base and future growth and development. They were defending and creating a market (van Someren and van Someren-Wang 2009). Copenhagen was the media hype for months. Copenhagen would save human beings from extinction. Two years later, it was silent, very silent about Durban. Durban is even more important than Copenhagen because the urge to act is even greater, and the Kyoto agreement is at stake. Suddenly, media but also policy makers lost interest. Of course the debt and euro crisis ask their attention and available time of policy makers. But in Durban, again the talks and negotiations are about creating a market for ecological industries. Who is playing the big silent role behind the scenes? Of course Green China. This time, the Chinese want to trade an agreement on a CO_2 cap for a transfer of environmental technology from the West (van Someren and van Someren-Wang 2011). In Durban, the "Realpolitik" is made not in Brussels.

In the past decade, the USA was lucky with the invention of shale gas and tear sands technology. Especially, shale gas is contributing to the realization of the aim of energy independency for the coming decades. But it is still a fossil energy source.

What about European industry and their signals toward the European Commission? Do they see the Green China picture? Also here, a representative study demonstrates the defensive strategy of the association of European industries (BusinessEurope 2011). Their key message is that "The EU and China depend on each other in the global economy, with Chinese and European companies being parts of fully integrated global supply chains…both the EU and China should increase their efforts to further expand the opportunities though stronger bilateral cooperation" and "Despite many large business opportunities a number of issues

Table 7.3 From triple P on paper to business

Sustainability issues	USA	EU	China
Ageing	Immigration keeps average age low	Pension age gradually lifted and immigration politically difficult	Family ties and start of social security and lifetime employment
Urbanization	No issue	Plans for eco-cities absent	Eco-cities source of future prosperity
Resources	Global sourcing of oil at center of policies	Strategic plan absent	Global sourcing of all natural resources
Mobility	Predominantly car industry driven	Predominantly car industry driven and weak policies	Electric vehicle program
Energy	Independency by developing shale gas and bio-based fuels	Failing implementation policies	Energy mix of fossil and nonfossil
Health	Obama security system	Expensive systems need reform	Communistic system and reforms
Social inequality	Tax reform	Expensive social security system	Go West strategy to involve other regions in wealth
Climate change	Climate change not as limit of economic growth	Aim of low-carbon economy	Climate change agreements as economic stimulation

make business operations with China challenging for European companies. Namely strongly state influence on the market, discriminatory treatment, forced technology transfer, subsidies, restrictions on raw materials and the fear of retaliatory measures." The report continues with a long list of measures China should take such as opening up markets, allowing foreign-ownership restrictions only for national defense, and increasing the recourse to international standards and agreements, like OECD provisions on export credits, and strengthen and simplify IPR enforcement, enhance transparency, and address concerns over retaliation. The report says the EU should react as a coherent block, develop means to protect its companies and member states from Chinese pressure and intimidation and implement a strategic raw materials policy. Of course, if the European lobby organization sends these signals to the European Commission as the main issues in the business relation with China, one cannot expect from the European Commission to send totally different signals to China. Most members of the European Commission and the civil servant apparatus have a pure political background and largely lack experiences in industry, private business, or innovation development. The EU and the private industry have to do homework in topics such as institutional innovation and sustainability enabling the EU to act as a fortress.

Table 7.3 presents some other examples of sustainability issues as driver for economic growth and development.

Table 7.4 How to improve the governance to win the sustainability challenge

Governance level	USA	EU	China
Worldwide	From military governance to economic governance of power and China strategy needed	Strategies needed for China, resources security and energy	Applying Triple P criteria instead of profit on their global sourcing strategies
Cross national	Agreements with neighbor states	High potential for pan-European cross nation state policies	Power of China hampers cross nation cooperation
National	Lack of national sustainability policy	Member state policies need at least more coordination	Five-year plans need midterm reviews to adapt to new circumstances
Regional	Regional economics underdeveloped	Regional economics underdeveloped	Provinces need complementary activities instead of copy–paste programs
Local	Local economies often too small for separate policy	Local economies often too small for separate policy	Size of local economies or megacities allow separate policies
Symbiosis between private and public	New governmental role needed to support new industries firms	New governmental role needed to support new industries firms	China fosters both state-owned firms and private firms

The core principle is that the Chinese policies try to transform most of the sustainable issues from limitations of current growth curve into an opportunity for further growth and development. In this way, paper policies are becoming practical business development areas and the source of strategic innovation.

7.8 Governance

For Chinese, Europe is one region. But there is no integrated Europe yet. The Chinese know. That is why they go business shopping in Europe on their trips. Therefore, national and local policies have to be mentioned as well. Europeans see Europe as a patchwork of different cultures and nations. They are proud on their own region and on the differences with fellow European regions. But for doing business with Chinese, it is better to use the Chinese vision; otherwise, they will continue their successful trick of playing one country against the other. Already only for this reason, European policy is very important.

In this respect, we should make a distinction between five levels. The first is the worldwide level, the second is the cross-national level, the third is the European level, the fourth is the level of national governments, and the fifth is the local government including provinces and municipalities. On almost all levels, a radical change is needed to countervail Green China. Why? Because we have seen in previous chapters that individual companies, regions, and the EU are missing opportunities partly because of failing governmental behavior toward China (Table 7.4).

7.8.1 The World Level

The biggest challenge and task for the USA is to abandon their military power strategy and to replace it by an economic power strategy. In our times, without a clear threat of one big enemy like in the Second World War, the military strategy is very ineffective.

The ineffectiveness is based on three issues. Firstly, it is ineffective from a military point of view because it proves hard to win from modern guerilla tactics and without a long-term commitment. The oil wars in Kuwait and Iraq may have secured future supplies, but the Chinese who were not part of the coalition forces are the free riders. Secondly, military operation is a very costly strategy consuming large parts of the governmental budget. Thirdly, human resources are allocated to a destructive industry, and post war traumas make the human resources not deployable anymore for the buildup of the future economy. Our purpose is not to advocate a weak military force but more to change power policies and the use of the military.

Green China asks for economic power. The past has shown that the economic power of individual entrepreneurship, market-based innovations, and the creation of power houses is a huge asset of the USA. This core competence of the USA will be needed in the future as well. Only the world order will be changed by Green China, and the dogma of free market economy or market-based innovations needs to be adapted to the upcoming new rules. One of them is to involve the government more than in the past without destroying the flow of bottom-up innovations and freethinking. The USA needs a China strategy and a new strategy for securing their resources. For example, the trend of insourcing of manufacturing companies requires a global resources security based on bilateral national agreements.

7.8.2 The Cross Nation Level

With regard to China, the cooperation with neighbor states will prove to be a difficult task. The size and power of China can be intimidating for smaller Asiatic neighbor states. But already now in several countries, China has outsourced some of its low-added value and low-profit margin manufacturing activities. For the USA, the cross nation cooperation boils down to Mexico and Canada. Mexico is still a low-cost country, but on the longer term, their wages will rise as well. Therefore, using neighbor states as low-labor cost sources is not a very smart strategy and postponing the necessary creation of a sustainable economy. The cross nation level is very important for the European Union.

In 1975, the relations between Europe and China were launched. Over years, the external relations were fed by numerous meetings, talks, summits, and high-level group gatherings.[2] The EU is the biggest trading partner of China which makes

[2] European Commission – External Relations, EU-China Relations: Chronology, http://ec.europa.eu/external_relations/china/index_en.htm

formal and good relations necessary. In 1985, the trade and cooperation agreement formalized these talks. This was followed by rounds of sectoral cooperation dialogues beginning on 2004. As of 2010, a next upgrade of the relation was agreed upon by a strategic partnership between the EU and China. On the agenda were topics like the global economy, climate change, and of course, the financial crisis.

The most clear illustration of failing implementation and its value conversion capacity is the Lisbon agenda. The Lisbon policy formulated in 2000 had the ambition to make the EU the most competitive region in the world. One of the instruments was an average 3 % of GDP should be spent on R&D. It appeared that this R&D ambition level was not attained, and becoming the most competitive region was also completely missed. Well, let us assume that bad things can happen in life and new chances will come to restore failures. This new chance is the climate policy Europe 20-20-20. In 2007, the European climate policy 20-20-20 was announced. This EU climate policy states that by 2020 the EU should use 20 % less energy, 20 % less climate gases, and a 20 % share of renewable energy in total energy use. Around 2010, it was admitted that probably these aims would not be achieved.

The EU also lacks a pan-European policy on sourcing natural resources. Only as of 2010, in the area of energy, the EU has been preparing an energy policy called Energy 2050. But the first unofficial concepts indicate that the centerpiece is aiming at lowering CO_2 levels by a new energy policy. Of course, lowering CO_2 is regarded to be important, but this is something else than aiming at an EU being independent or at least less dependent on other countries for their energy supply. Even better would be to strive for more bold targets like a fast buildup of renewable energy supply and the required smart grid infrastructure. Moreover, fostering the aim energy independency of foreign countries requires large investments in decentralized energy solutions. In Europe, the implementation of a smart gird already proves to be one of the biggest bottlenecks and requires governmental action.

These missed opportunities of giving the EU some right to exist give us the right to have some doubt about other policies. What should we think of the EU 2020 aiming at an innovation union? And what about the EU road map toward 2050 aiming at a low-carbon society and energy supply? Green China has also plans in this direction as the 12th year plan says. Do the Chinese fulfill the threefold criteria of formulation, implementation, and value conversion? Yes and no. No because execution of laws is a big problem in China, and central plans are often frustrated in the region. Yes because making big plans is a core competency of the central and local governments. Yes because implementation is not always achieved, but because of a strong link between government and industry, some plans are realized. Of course, this holds for very concrete and practical plans like building infrastructure and less for improving health or services. Execution and value conversion of plans can be seen in the wind and solar industries. The link between planners and executers enabled the Chinese to establish world players in wind and solar sector. They did so by creating a home market and domestic appliance followed by export to the USA and EU.

7.8.3 National States

With regard to national governmental planning, China shows the way. China has proved to transform their five year plans from communistic ideological plans without economic reality into nationwide strategic blueprints for sustainable growth and development. The past decade has shown that the five year plans have now economic sense and future industries have been created. But their governance of the five-year plans needs some improvement as well. The copy–paste and me-too behavior on the level of the provinces leads to the same output and overcapacity.

The USA lacks a nationwide approach for sustainable development. It is both the success of the market ideology in consumer markets and decentralized government on state level that now hinders a governmental sustainability approach. Like the Chinese silently abandoned their communistic plan economy, the USA should in certain cases and situations skip their market ideology.

In the situation of Europe, the nation state is again a complex one. It is about national identity, sovereignty, independency, and rivalry. Some European citizens feel like a European citizen, but most feel like being an Austrian or Italian. Some even feel more connected to their local environment around their hometown. For large parts of the population, localism dominates globalism. Giving up sovereignty to Brussels is politically difficult and not accepted by many Europeans. Being independent is still preferred but impossible with rise of Green China. Moreover, the European project is about cooperation. But rivalry about who is the real leading state, especially between Germany and France, is the actual battle behind the curtains. But China and the USA are not interested in these rearguard fights.

National European policies have two faces. The first is that national policies sometimes cannibalize or hinder European policies such as in energy issues.. The second is that sometimes, national policies can be successful such as French nuclear industry is supported by French government. But the latter cases are rare. Even after many years of bilateral meetings, exchange programs for civil servants, learning from each other's projects, and so on, European governments still have great problems with defining their role in setting up and governing the relation with China over a longer period of time. This is not only caused by a lack of knowledge of Europeans but also by a lack of knowledge and understanding at the Chinese side how Europe works. Anyhow, if we want to do business in China, we have to adapt to their habits and business practices for a great deal without losing our own face and dignity. Apart from illegal practices and similar not allowed behaviors, there is a lot which has to be changed as we have seen in earlier chapters.

Most member states have bilateral relations on political and policy levels. They cooperate with the central government, and most ministries have a working relation with each other. For the Chinese, the politicians are still much more important than the businessmen. Chinese central government gives the march order, and the state-owned enterprises follow and carry out. Although the private companies have no direct link to central or local government, they still have to deal with government about their role and freedom to operate. Therefore, European private business has to understand and in particular, to learn how to fine-tune with their own government to realize their business aims related to China.

To our experience, most firms are still not aware how to play this game. In Europe the government and the private business, are used to completely other forms of co-existence and cooperation than what is needed in China. Of course, in Europe, differences exist. The French government is traditionally more centrally organized, and some companies and industries are more politically governed than others in Europe. This French mercantilism fits better to the Chinese situation. But Germany shows that French mercantilism is not the only road to success. The Germans are less centrally organized than France.

7.8.4 Local or Regional Governments

Local governments also have bilateral relations with their official counterparts. Within Europe, there is no standardized political organization on the local level. But in a rough way, the level of provinces and municipalities do exist in one or another form, and they have their counterpart in China. But often, the dimensions differ. For example, some German provinces (Länder) have 25 million inhabitants and Chinese 80 million or more. It is the same of course for the cities. But the difference in size is not important because even European cities and provinces still have to offer something. For example, from the Emilia–Romagna region in Northern Italy, the Chinese are eager to know how family-owned private small businesses can be so innovative and successful.

Like on the national level, there is a lot of traveling back and forth to China. Delegations come to local regions as well. But despite the formal economic agenda and program, very often the complete Chinese delegations or some members are not at all interested in cooperation. Sometimes they make an allowed touristic trip to Europe under cover of an official visit.

These activities are nice trips but the core question for governments 'What is our vision, aim and attitude on China?' remains often unanswered. This will determine how the bilateral relations and discussions between the two develop. Here comes the issue of cross-cultural management into play.

7.8.5 Symbiosis Between Private and Public Organizations

Another missed point is the involvement of the government. From both central and local government. The trick is not only to find the right decision makers, which can be hard enough, but to use those relations not for penetrating the market but for developing the market on the long term. Why? From the penetration phase, nobody can earn money. Money can only be earned from established and growing business with enough volume measured as scale or as luxury or specialized niche market. For this reason, Chinese prefer pilot projects. They have two goals with that. See if it works and try to get your knowledge. The Dutch government has introduced a very interesting idea with regard to the core role of their embassies. In 2011, it was decided to focus the embassy activity in supporting and promoting Dutch business instead of a large scope of activities. Some criticize this very narrow role of diplomats as economic tradesmen. But for a small country, it might be an extra weapon, and at least in China, it will certainly help.

You involve government on both sides also during the growth phase even when you think they are of no use anymore. Why? When ministers or high-ranked officials back your business, it is for Chinese much more difficult to retreat from the project or cheat the business partners. But sometimes, they have no scruples at all and still dump you. Well, that is a business risk you know on beforehand, so do not be surprised and take your loss as a man. And start over or quit.

7.8.6 Cross-Cultural Management

The issue of cross-cultural management can best be explained by an example. Our example here will be behaving and acting from moral superiority or strength. First, moral superiority. On all three governmental levels, European governments act from a conscious or unaware moral superiority over others. Europe can be truly proud on their achievements to have been able to establish a more or less free market economy. But it took over 2000 years. And the European project is not yet finished as the euro crisis has shown. Even worse, years ago, the issues of human rights and Taiwan were high on the agenda because after the cold war, the capitalist system had conquered the world and with that, the moral values. But the financial debt and euro crisis showed that our system is not that all ideal. Moreover, other norms and values also have a place in the new world. Not only Asian ones but from Islam as well. The difficult relation with Turkey shows this very clearly. This attitude might be described as moral superiority. But in Chinese eyes, any feeling of superiority must also be shown on the battlefield.

The nowadays battlefield is the economy. Exactly here, the view of the Europeans of themselves and the Chinese differ. Especially in the last decade, a big U-turn has occurred. Whereas the Chinese economy has grown fast, the European economy went into trouble. The problems of ageing, immigration, low growth rates, and failing integration, to name a few, slowed down Europe's growth and development.

The Chinese admire some of Europe's achievements like luxury brands such as Vuitton, technology and environmental laws. But deep inside, they also believe Europe's time is over, and it will become a museum. The reason is that Europe talks with 27 heads despite a European president and high representative of the Union for Foreign Affairs and Security Policy. Europe is not a superpower but a divided old world. And it acts like that. For the Chinese, this is fine. For the Chinese, Europe is a lucky dip of technology. They love to play one against the other. It is all part of trading and negotiation skills. They love these games. Especially with foreigners. Europeans are foreigners. Some with a bad history. Now, it is payback time. Look at the Bund in Shanghai, on almost every house build by the Europeans, now wave

Chinese flags. A sign of power and superiority. And showing who is the boss. The Chinese are in their own eyes superior to any other race and region in the world. The lack of power can be seen in the failing energy policy, the failing natural resource policy, and inadequate innovation policy. In dealing with China, Europe should become a powerhouse and act from strength not from moral superiority. The only conclusion we can draw is that Europe has first to do some homework before trying to deal with China.

Now, we deal with strength. The strength of the EU can be measured in total GDP, income levels, population, and other indicators. But these indicators only say something of the past strength not of the future power. As we explained earlier, the future battlefield is the sustainable economy which requires formulation and implementation of strategic innovations. Therefore, from a governmental perspective, future strength can be measured in the ability to create and implement policies and above all to measure their bottom line impact on their contribution to Triple P value conversion. How does the EU perform on this threefold criteria package of formulation, implementation, and value conversion? Almost every government or governmental institution is able to produce paper policies. There is no doubt about that. But already with the second criterion, most European governmental institutions will come into trouble. Not only the implementation fails too often also law enforcement is absent or unexpected results do occur. The evaluation on value conversion or a direct link to value-added activities by involved stakeholders is a blind spot in the world of European policy makers. Let us take a few relevant examples which can match China's policies or which should be executed if the EU takes itself serious.

7.9 From Defense to Offense

This chapter demonstrated in a nutshell that neither the current EU nor the USA is able to protect national economic interests with defensive policies against China. Behind the scenes, there is a lot of angriness and frustration, and bottom line is, there is lot of red ink. But to the defense of governmental and business leaders, we must say that no standard solution exists. And the Chinese seems to be undefeatable because we try to win with our weakness. The Chinese are human beings as well making mistakes and wrong decisions. But you have to find their weak and vulnerable points compared to the West. The Chinese are masters in hiding the weaknesses and in turning weakness into strength. They learned that from Sun Tzu. The West knows too little about the real China. What we learn about China is most of the time filtered by Western China experts who can only present the Western interpretation of China. But that takes some effort and time to find out. We spend much too little time for self-reflection and even less time to find out who the counterpart, what his background is, who are in his or her network, and what relations and reputation they have.

There is only one conclusion possible. Radical renewal of Western China policy is necessary. It is useless to ask and beg for more openness and transparency. It is also useless to build a defense line as both the European Commission and the industry representatives do. The Chinese will respond by saying that on crucial issues and turning points of the EU, transparency and democracy was put aside for the blessing of the European people. Remember the introduction of the euro, the European Constitution, and the measures taken for fighting the financial crisis. They will continue the relation with the EU with the arguments that China is at a turning point as well and China is building a harmonious society for their own people. But China wants to live together with other nations in a peaceful world. What they really mean but not say is: China first. Only an offensive strategic response can countervail Green China. But then, Western policies and instruments have to be radically changed. We need governmental strategic innovation.

But which direction should the Western government try to innovate? Not by working against the private sector. Only a combined approach between government and private sector is able to counterattack Green China. This will be elaborated in the last chapter.

References

Barroso, J. M. D. (2011). European renewal. State of the Union, European parliament, Strasbourg, 28 Sept 2011.

BusinessEurope. (2011). Rising to the challenge. October 2011.

de Gucht, K. (2011, November 8). Looking beyond the crisis: Making the EU-China trade relations work, EU-China high level political forum, Brussels, speech/11/728.

Geithner, T. F. (2010). Treasury Secretary Timothy F. Geithner Testimony before the Senate Finance Committee, Press release TG-740, 06.10.2010.

Someren, T. C. R. van (1995, January–March). Sustainable development and the firm: Organisational innovations and environmental strategy. In R. Welford (Ed.), *Business Strategy and the Environment* (vol. 4, Issue 1, pp. 23–33). John Wiley and Sons.

Someren, T. C. R. van, & van Someren-Wang, S. (2009e, December 9). Klimaattop? Welnee, er wordt een markt verdeeld. In: *Trouw* (p. 29).

Someren, T. C. R. van, & van Someren-Wang, S. (2011e, November 30). China speelt in coulissen grote rol in Durban. In: *Trouw* (p. 19).

Someren, T. C. R. van, (1998a). Strategies for sustainable profits. In: *Milieumanagement van Kosten naar Baten*. Wetenschappelijk Instituut voor Milieumanagement, Universiteit van Amsterdam, 7th April 1998, Amsterdam, The Netherlands.

Someren, T. C. R. van, (1998b). Strategies for sustainable profits. In: *International BMW Seminar*, Dietramszell, 17 Sept 1998.

U.S.- China Strategic and Economic Dialogue. (2011a). Remarks by Secretary Tim Geithner at a Discussion on the Upcoming U.S.-China Strategic and Economic Dialogue, 5/3/2011.

U.S.-China Strategic and Economic Dialogue. (2011b). U.S. Fact Sheet – Economic Track, 5 Oct 2011.

US Consensus Bureau. (2012). *Trade in goods with China*. http://www.census.gov/foreign-trade/balance/c5700.html.

Wei, Z. (2011). *Global economic outlook grim*. Reuters: China tells U.S. trade talks. 21 Nov 2011.

White House. (2011). A strategy for American innovation: driving towards sustainable growth and quality jobs, 2009 updated 4 Feb 2011.

6th EU-China Business Summit. (2010). Innovation as a win-win solution for steady growth. Joint Business Declaration Brussels, 6 Oct 2010.

8 The Road to Sustainable Coexistence of China, the USA, and the EU

Different Directions for Future Economic Growth and Development

Green China is a real challenge for the USA and the EU. It is not the rise of China itself that causes a change of course but the rise of a total new world. In this new world, Green China has the ambition to become the number one. Moreover, the new world is not the current world added up with China and some other BRIC countries. The new world is accompanied by the introduction of new technologies, new organizational forms, new performance measures, new institutions, and new cultures, and sustainability is the leading principle of them all. Both sustainability and Green China are a worldwide issue. This implies that it will take three to tango. The USA and Europe have to define their new role in the new world. It is useless to point finger to China what they have to do. It is time for the West to take action and start with strategic innovations in public and private sector in order to become a right dance partner of Green China. But China is also at the beginning of a new era and needs to focus on strategic innovation in order to realize their harmonious society as a Green China society.

8.1 Crush or Create?

Green China challenges the West in every aspect. At first sight, the greening of China looks like the adoption of sustainability principles. Looking closer, sustainability has been chosen by the Chinese as a lead issue. Contrary to the West, sustainability has not primarily been chosen because of creating a better world. Sustainability is the core theme because it represents the next long-term growth curve enabling the Chinese to emulate the West in the field of technology and economic power. But the long-term strategic aims of China are a challenge for China itself as well. It is not a matter of time till China will be indeed the number one super power of the world. Therefore, a tri-Star world consisting of China, the USA and the EU is possible. But all three have to work very hard to stay on an equal level. Therefore, Green China is not about ecology but about the real motive of

T.C.R. van Someren and S. van Someren-Wang, *Green China*,
Management for Professionals, DOI 10.1007/978-3-642-28810-4_8,

becoming number one and its accompanying behavior. Chinese green is not identical to Western green washing but a natural disguise of the dragon. A dragon should be green. But modern dragons can change colors and show flexible behavior which make them dangerous but not unbeatable. At least, this is the effect we want to achieve with our book title. Western readers expect a book about ecology or sustainability in which environmental issues are central. Chinese readers know about the different color shades in their society. There is always another shade which represents the real motives and aims.

The past decades were the most easiest time for China, and the country is now entering the most difficult and challenging time of its modern history. The Chinese box of Pandora has been opened, and now, the leaders have to cope with numerous issues. The greatest challenge for Chinese leaders will not be the development of technical innovations but the nontechnical innovations fostering the government's target of indigenous innovation. Here, the necessary conditions of freedom, bottom-up innovation processes, independency of higher power, and changing social conditions free the energy of the Chinese people. Governing and controlling the endless human energy will be more important than fossil energy and renewables.

Although the mutual dependency between the West and China has increased, the future of the EU and the USA is not completely dependent on China but on the necessary strategic innovations to be initiated and implemented in the Western home turf. That is Western homework for the coming years. Furthermore, the West needs to get a real understanding of Green China. The understanding should go beyond cooperation agreements and focus on Chinese culture, social, and strategic behavior. Only then a proper China strategy to encounter Green China is possible.

China has always two faces. In Chinese eyes, there are no single equivocal situations. In Western eyes, there must be a single truth or answer. These opposite mind-sets are the root cause that forms many misunderstanding and developments paths for the future. China has become the biggest emitter of green house gases but is also the biggest manufacturer of solar panels. You cooperate with Chinese businessmen for years, and the next day, they dump you and set up the same business with a family member. Till recently, polluting non-Chinese manufacturers were warmly welcomed to invest in China; now, most of them have much more difficulty to start business. Is it arbitrariness? No, new situation emerges which requires new measures, attitude, and behavior. Sometimes it just happens like financial crisis. Sometimes it is planned for, like developing green industry and Chinese global green powerhouses. Our message here is that Green China is not only a threat for the Western way of life and society or only an opportunity. Green China is both.

Be not blinded by the fast growth of China and their fast decision-making which makes many Western business leaders and politician sometimes jealous. Also, the Chinese know that there is always another side. But it would be a mistake to think that the downside of fast growth and development will slow down Green China automatically or from inside. Green China needs to be challenged; otherwise, Europe and the USA will be wiped out. The West has to prepare for Green China as a threat as well as an opportunity. In both cases, a new Europe has to be created, and the USA has to reinvent itself into a sustainable nation.

In the previous chapters, the threats from Green China have become clear. The several crises elevated the financial sector from all the other industries as a system relevant industry and the most important one. Almost all available money went into the financial sector to save banks and currencies. And still, at the end of 2011, Europe had to beg for money from the BRIC countries. But it demonstrates that European senior officials did not understand the position of China (van Someren and Van Someren-Wang 2011). China refused because first, Europe has to do their homework to restore financial institutions and governmental debts, and secondly, the risk of not getting back the financial inlays was too high. The latter shows that Europe needs industries with future earning power. The growing power of, for example, Chinese wind, solar, and battery industries shows where the industrial threat is coming from. This creative destruction is part of the normal market process. The Chinese threat comes in a new form. It is not about a new business model but about new strategic innovations. Green China industries are supported by five-year plans and governmental support on the individual firm level. Simultaneously, private companies without direct governmental support also are part of newcomers. The West encounters two faces which both form a strength on their own.

The biggest internal threat comes from misinterpretation of Green China, uncoordinated Western policy making, short-term strategy, selfish interests, and absence of implementation power. The positive interpretation would be to see it as competition making the West stronger and more innovative. But with regard to sustainability, it has been demonstrated that in certain sectors, the dimensions of required infrastructure, investments, new competences, and long-term timelines go beyond the possibilities and resources of individual firms. Individual companies can introduce new telecommunication technology like mobile phones or even sustainable chocolate bars. But for sectors like energy or electrical or hydrogen cars, a new very expensive complementary infrastructure is necessary. Moreover, the economic engine of renewable energies takes a long time before they are cheaper than existing fossil fuel-based energy system. At the end, the economic engine, or the cost factor of Triple P, decides about acceptance and breakthrough and growth of renewables. Some form of government combined with private industry is needed to bridge the long time gap. The private industry together with knowledge institutes is needed to create and develop new renewable systems with competitive economic engines.

The ability to create innovations had been part of European heritage, but the last decade, somehow, a hiccup delays and lowers the innovation output. The hiccup consists of the absence of an institutionalized innovation system in the majority of private firms. Another hiccup is the failing quadruple helix structure. It is not only the cooperation between private, public, and knowledge organizations, the triple helix, but the financial support which bridges long-term time gaps and fuels the growth of companies. The latter is known as the innovation trap. However, the government has the responsibility to support to create a market for sectors like renewable energy and electrical cars. The slow growth, high investments, and initial losses of electrical cars cannot be financed by individual firms. Therefore, cross nation home markets need to be created. This is part of the new symbiosis between private and public sector. It is insufficient for governments to make available money such as the feed in tariff in Germany. This is traditional division of private and public tasks. The German solar industry has a Pyrrhus victory because the Chinese solar panel manufacturers flood the German market. The Chinese earn money from the German tax payer which will be used to wipe out German manufacturers and take them over to get technical knowledge. Here, Europeans have to get streetwise. But in every situation, different and unique solutions have to be created.

Pan-European policies instead of national polices on industrial development and greening of industry are needed to support the creation of European green powerhouses and make use of Europe's diversity. Some companies show the green path of growth like American IBM, German Siemens, and Anglo-Dutch Unilever. These companies focus their strategies on the green wave and megatrend independent of policies or whether or not protocols in climate conferences like Durban are agreed upon. But still, we also showed that competing with Green China needs more than a green strategy riding the waves of green trends. The threat of insufficiently taking into account the cultural aspects of dealing with Chinese government, business partners, and customers can be more detrimental than anything else.

The opportunities of Green China are the tremendous problems with sustainability. Green China has an increasing demand for green solutions in almost every industry. Green solutions are not identical to technological solutions. Instead, a combination of technical and nontechnical solutions will offer the ideal combination of instruments to realize Green China. Particularly, the non-technological aspects of strategic innovation are the weakness of Chinese political and business leaders. Their bias for technology offers the greatest opportunity and challenge for Europeans. The creation of necessary strategic innovations in the area of leadership, international business development, sustainable companies, and institutional innovation requires more than a green technology, a good governmental policy, or cheap capital.

Shall we sacrifice the free market principle when aiming at a new symbiosis between public and private organizations? No! For several reasons. First, the complete free market is a myth. Oil companies have strong links with governments; in Austria, the notary sector is a protected market by means of regional monopolies, the health sector is a highly regulated market, European energy markets are highly regulated, and subsidies distort many free markets. Second, we do not advocate government-owned companies. Instead, governmental support and commitment can be crucial for the long-term success of private sector while dealing with China and sustainability. This support creates the right conditions for strategic innovation and should go beyond opening doors. Western governments can and should play a central role without direct investment, subsidies, or ownership. Western governments should be part of the total power play between governments and private and publicly held companies. Chinese governments need a countervailing power beyond defending illusionary free market principles. Even the biggest multinational cannot be left alone in doing business with China.

8.2 How the USA, Europe, and China Can Reinvent Themselves to Become Triple Stars

Due to the trade volumes and current complementary activities of basically manufacturing and all the other value circle activities, China, the USA, and Europe are like triple stars circling around each other. For astronomers, it is possible to predict the trajectory of two celestial bodies. Two celestial bodies move around each other in an elliptical orbit. But as soon as a third body is added, it is much harder or impossible to predict accurately the motion. Economists cannot predict at all, so how should economists predict the development between the two earthly bodies Europe and China or even three when the USA is included? And what about the rise of India, Brazil, and Russia? Of course, physical laws cannot be compared with economic laws. But one cannot and should not predict.

However, what is needed is the development and growth of Europe and USA; otherwise, the motions of the Western bodies will be dominated by the Chinese one. Instead of making projections and predictions about future role of the Chinese economy and society, the priority should be on the sustainable development and growth of Europe. Otherwise, the European Union will end in a supernova watched by fireworks-loving Chinese and the rest of the world. We will not be missed when looking at the attention paid in American and Chinese media to Europe. So the EU has to save itself by reinventing itself. But then we need visionary politics and leaders without fear for daily polls or their job. Politicians say Europe needs entrepreneurs. They are right. But Europe also needs world leaders with strategic foresight.

Around 2010, the USA, European Union, and China are dependent on each other when looking at import–export statistics. China lives from export to the European Union and the USA. But be aware it is not a static system. China moves in the direction of an innovative economy but needs to pay attention to social insecurity and ecological issues. Both the European Union and the USA have to pay attention to reinventing their sustainable economic engine and green earning capacity and define new social core values needed to define its reason of existence and its role in the future world.

The European Union and the USA have to formulate a grand plan to grasp the opportunities of future earth spaceship. But also, China needs another leap forward. These strategic innovations grand plans have to be formulated around the themes mentioned respectively in Figs. 8.1, 8.2, and 8.3 and Tables 8.1, 8.2, and 8.3. The next sections shortly indicate the most important issues.

8.3 Grand Plan for the EU

The EU is flooded with plans by the European Commission and the governments and policy makers of the member states itself. Therefore, it is useless to add another plan but instead cut the crap and focus on the key issues for creating a countervailing

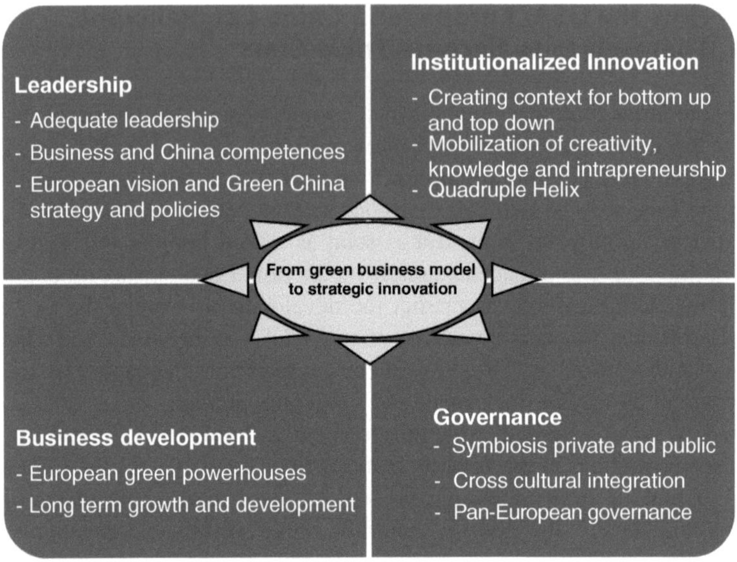

Fig. 8.1 Issues for strategic innovations in the EU

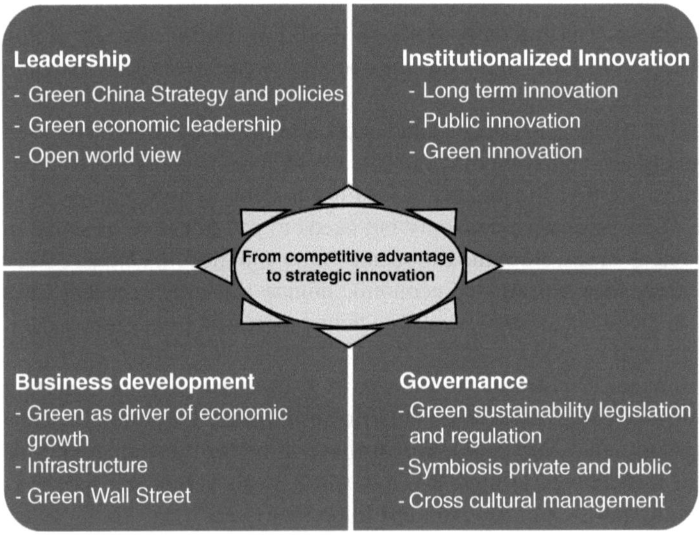

Fig. 8.2 Issues for strategic innovations in the USA

power to Green China. In Fig. 8.1 and Table 8.1, the most important issues are presented.

Figure 8.1 gives the diagnosis and Table 8.1 the direction of the solution. The message from Fig. 8.1 and Table 8.2 is the answer to Green China renewal of the European Union project. It starts with leadership. Currently, the European Union is

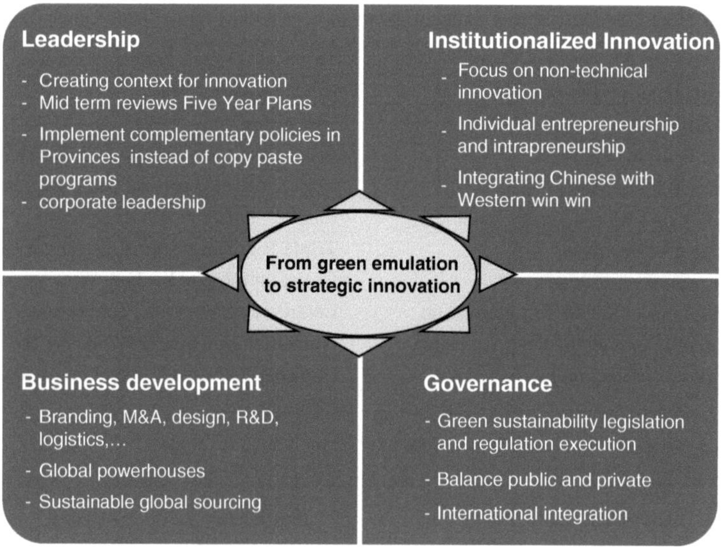

Fig. 8.3 Issues for strategic innovations in China

Table 8.1 The EU direction of strategic innovation as a reaction to Green China (examples)

From	To
Fragmented leadership	Pan-European vision and China strategy
Lifetime bureaucrats	Business experienced and China expertise
Ecology-driven sustainability	Triple P-driven sustainability
Knowledge economy	Entrepreneurial economy
Slow growth SMEs	European green powerhouses
Subsidy forest	Green investment banking
Short-term election-driven changing policies	Long-term commitment
Member states institutional boundaries	Pan-European governance
Culture independency	Cross-culture management
Technology content-driven	Context-driven innovation

without a vision about the rise of Green China and how the EU should react and act. Part of the cause is that the large majority of European and national governmental leaders lack significant knowledge and experience with private business and dealing with China.

Too often, European careers in government are based on party politics and lifetime civil service. How can policy makers formulate good business support policies when they lack hands on business experience? In Europe, the task of the government is to create favorable conditions for sustainable growth and development and not taking market decisions or playing the role of entrepreneur. This is a defendable position, but it requires an extremely important precondition. Paradoxically, the separation between market and government sets highest requirements for

Table 8.2 The USA direction of strategic innovation as a reaction to Green China (examples)

From	To
Inward American perspective	Outward world view
Military leadership	Green economic leadership
Trade and IP-rights conflicts	Insight and understanding Chinese strategy
Market-based innovation	Market-oriented private and public innovation
Dependency on Chinese money lending	Earning money from Green China
Ecology against economy	Triple P-driven sustainable growth
Short-term money machines	Green investment banking
Short-term bottom line-driven innovation	Long-term green value conversion
Education of talents in the USA for China	Utilization of Chinese talents for the USA
Environmental protection law jungle	Comprehensive sustainable law system

Table 8.3 Direction of strategic innovation for Green China (examples)

From	To
Economic Chinese leadership	Sustainable world leadership
Decisive top-down one-man show	Organized 1,000 flowers blossoming
People/network management in China	Customer/market demand in the West
Management by trust	Management by competence
Technical innovation	Nontechnical innovation
Vertical innovation process	Vertical and horizontal innovation process
Private business development by gambling	Strategic innovation management
Grand plan-driven M&A acquisition	Post-merger integration plans
Green capital accumulation	Risk management and allocation
Paper tiger law system	Integrated and control-based law system
Environmental protection law jungle	Comprehensive sustainable law system

governmental leaders. The necessary switch from technology-driven innovation and governmental decision taking toward creating the right circumstances requires excellent knowledge about markets and market behavior. This condition is not fulfilled. Nowadays, this omission leads to comitology, lobbying, growing forest of subsidies, and often, contradictory regulations.

Particularly, the combination of sustainability with the rise of Green China requires industrial leadership and knowledge of doing business with the Chinese. This is reinforced by the fragmented leadership within the EU leading to protecting own national interest. The minimal best outcome for all is the compromise, whereas the maximum outcome for the EU is missed.

In the European Union, not only a gap between politics and civil society has been created but also a huge discrepancy between private and public sector exists. The consequence is, for example, miscommunication and inadequate policies, hindering the foundation and growth of start-ups and SMEs, innovation policies, and transformation to a sustainable society without clear success.

Sustainability requires government creating new conditions and closing gaps in capital and time. The famous Death Valley is not solely a finance problem.

Government can create much more assurance in markets when long-term policies beyond political election periods are at the core. Instead of subsidies, a switch to green investment banking can prove to be more successful and rewarding. The EU spends hundreds of billions to save the euro but forgets to invest in future earning capacity of the euro zone. It should be the other way round: invest in future green industries which will lead to a stronger economy and future prosperity. Only then can an own currency make sense which need not be saved.

The role of government is not a copy–paste from China. In the Western view, the government is not a good entrepreneur. But government can compensate this competence by market insight and long-term policies fostering the greening of industry. Environmental laws and policies in Europe are focused on the detrimental side of economic development and corporate behavior instead of creating new industries and new firms based on Triple P. The combination of stable long-term policies and industrial development fosters green innovative entrepreneurialism.

Business flourishes in circumstances of freedom and bottom-up initiatives. The strength of the EU is to mobilize entrepreneurship, creativity, and knowledge. Moreover, the diversity within Europe is now a weakness because we do not exploit it for fostering green innovations. The establishment of European green powerhouses is necessary to counterattack the globalization of the Green Dragons. Europeans do not win from China because of lack of green powerhouses. The Chinese have more respect for the bigger companies than for smaller companies. These powerhouses are needed to conquer global markets and to make the Green Dragons a difficult life. At the same time, these powerhouses can exploit the biggest potential asset of European business: the bottom-up innovation processes. The latter still has a huge potential for improvement, and the context-driven innovation is needed to create a sustainable society. This is the main task of business leaders.

The symbiosis between private business and public sector is in the Western market context a difficult but necessary one. With regard to Green China, businesses and governments should team up. The symbiosis does not mean that a government should take the market decisions in which industry to invest or not. The governments, however, can support private industries by building up relations with their Chinese counterparts. Such long-term relations help to support private business and make Chinese government more difficult to take decisions against the interests of private business. In Europe, it is still underestimated what important role is central but even more local governments play. Particularly, how to play the Chinese game still needs improvement.

Cross-cultural management includes doing business with Chinese but also using in a right way the available Chinese talents. In Europe, very often it is thought that Chinese talent equals Chinese grown-up in Europe speaking the language and knowing the habits or Western Sinologists with business experience in China. In Chinese eyes, the real talents are those who have grown up in China, have been educated there, and gathered some business experience. The reason that they are the real Chinese is not because of their presence in China but because of their network. Taken together, the early lifetime friends, school- and study mates, military life and working life combined with knowledge, and working experience make the real treasure talents. The West needs these kinds of talents to grasp the opportunities and steer between the black holes and golden mountains.

Based on the issues in Fig. 8.1 and Table 8.2, the strategic innovation agenda for the European Union may be as follows:

1. Vision of a long-term sustainable EU 2030 in a future tri-Star world with China and the USA.
2. Reinvent government to cope with the sustainability growth curve and Green China.
3. Formulate a Green China strategy and the role of the USA. Combine the USA and the EU as a Western home market and revive transatlantic relations. Formulate a global sourcing strategy for natural resources.
4. Create or set conditions for pan-European powerhouses.
5. Symbiosis of private and public organizations to realize long-term development of green technology and infrastructure.
6. Foster cross border regions instead of nation states and exploit diversity in people, knowledge, and creativity.
7. Foster quadruple helix approach in which budgets are reallocated from social security to strategic sustainable sectors.
8. Cooperate and compete with Green China based on sustainable win–win instead of technology for market trade-off. Apply symbiosis to create a countervailing power against Green China and to enforce a mutual win–win.
9. Integrate Chinese sea turtles (talents oversea) and bind them in Europe and use them as link to Green China.
10. Foster institutionalized innovation within corporations and exploit the mobilization of knowledge, intrapreneurship, and creativity.

This 10-point plan needs a new context. The biggest challenge for the EU and its member states is to complement leaders in Brussels and national governments with hands-on business experience instead of lawyers and lifetime civil servants without market experience. The dominance of technical knowledge of Chinese leaders needs also a strategic answer. Due to lacking business experience and listening to lobby groups, the Pavlov reaction of the current leaders is to create new rules and govern by law and regulations. Europe needs streetwise visionary leaders who are able to support private business and no apparatchiks. Only then can Green China be challenged by the 10-point plan. The aim is not to stop Green China or to maintain status quo of the West being number one but to ensure a place near to future Green China.

8.4 Grand Plan for the USA

The size of the USA combined with the implosion of the former contender Soviet Union allowed them to have an inward perspective. The rise of Green China has to invert this behavior. In Fig. 8.2 and Table 8.2, the main issues for the USA are addressed.

The US economy is still one of the most innovative market places in the world measured along R&D expenditures, innovation output of firms, the crucial role of risk capital, and research institutes. The triple helix or even quadruple helix is working quite well. At the same time, it is amazing to see that the new growth wave of sustainability has long been denied or ignored by politicians and private sector. Fast-moving short-term products like chips and iPhones are the standard to beat, but slow-moving long-term opportunities like renewables do not have the same innovative reputation. Both the government and the private business need to reinvent themselves.

Both the government and private sector need a long-term vision on the new growth curve and building up a relation with China on a new fundamental basis. The current battles on several issues like currencies and trade deficit belongs to the behavior of the old growth curve and not of the new green growth curve. The USA needs insight in the background and behavior of China instead of pushing American views. American values will not disappear, but they require a new context to stay as powerful as before. Like it or not, the rise of Green China and its increasing power requires at least understanding of the potential new super power.

Sustainability needs government, and this is the main reason for the USA to lag behind. A future green economy can be supported and triggered by new national laws and regulations. Green China has started to conquer worldwide markets including the USA. The USA needs an outward looking view on changing world order, and a China strategy is needed to reposition American (economic) interests. The Chinese attack the USA in the heart of their society: the penetration of the first phase of a new growth curve. Investment in green economic leadership instead of military power is the solution. The market-based innovation strategy of the current American administration needs to be complemented by creating a national sustainable framework of laws and regulations, a national green deal for investment in public infrastructure such as smart grids. The Triple P-driven growth perspective offer opportunities for green investment banking and private industries.

A brain drain combined with entrepreneurial spirit from East to West will hit China harder than the intensified US military presence in the Pacific since 2011. The USA has the tradition of being a melting pot of different cultures. This trend will continue in the future. Chinese talents should not only be given the opportunity to finish study at famous American institutes but they should be kept for building up the American new growth curve. West meets East gets a new perspective: America's West coast is the new neighbor of prosperous East coast of China. The USA should formulate an irresistible go and stay East program for Chinese.

Some US firms are heavily involved in Chinese market, but they lack sufficient governmental support for surviving in the long term. The governmental support should switch from trade discussions and level playing fields to committing Chinese local and central governments to American interests. This sounds impossible but it is not. The strengths of some American firms in, for example, ICT support the will and desires of any people for freedom of applying knowledge into entrepreneurial success. American values can gain power if they are decoupled from military power. At the end, the power of the people is stronger than any other power.

This is the paradox of the People's Republic of China and America's biggest asset. The green economy can be built upon this old American tradition.

Based on the issues in Fig. 8.2 and Table 8.2, the strategic innovation agenda for the USA can be as follows:

1. Make a transformation from a military-economic superpower into a Triple P superpower.
2. Reinvent government to cope with the sustainable growth curve and Green China.
3. Abandon China as a rival policy and use power and resources to rebuild the USA as the new promised land.
4. Foster a nationwide grand sustainability plan and prepare government-coordinated infrastructure. Foster global sourcing resources strategy based on Triple P values.
5. Formulate nationwide strategic sustainability laws and regulation to stimulate both new green industries and decrease environmental detrimental effects.
6. Abandon ideological market-based innovation in areas of long-term green industry buildup and introduce governmental supported time gap bridging strategic innovations and infrastructure.
7. Create and foster conditions for green Silicon Valleys.
8. Transform Wall Street in green Wall Street to reallocate capital from financial and military complex into green industries.
9. Create new green powerhouses based on institutionalized innovation and long-term strategies fitting to green Wall Street.
10. Transform short-term market-based innovation in long-term private–public market development and creation of new industrial base.

This 10-point plan also needs a new context. The USA has two main challenges. First, reallocate resources from the military complex to the green tech sector and use power to counterattack Green China on their core strength of large-scale economic power. The innovative power of the USA can still beat the Chinese scale strategy. When worldwide military power relations with other nations are replaced by economic power relations, the involved parties will all gain. China has great respect for the strength of the US forces, but they are not afraid of it. China is more afraid of the innovative economic power and its human creative power. Chinese admire Steve Jobs giving them technical gadgets allowing individual freedom. The unavoidable next step to any kind of collective freedom really scares many political leaders. This is the real power of the USA. Secondly, the core idea of a small government makes in the American context sense. But with the rise of Green China, this mantra has to be reformulated. Not a bigger government is necessary but a government supporting the industries of the new growth curve.

8.5 Grand Plan for China

The rise of Green China paradoxically forces China to reinvent itself. In Fig. 8.3 and Table 8.3, the main issues for strategic renewal are summed up.

The Chinese success formula is long-term central planning combined with large-scale investment. This is very vulnerable because the created power houses took advantage of the green technologies mostly created in the West. For the Chinese, this market success is only temporarily. Without their own breakthrough strategic innovation, once the low cost-scale advantages have gone, the Chinese will be in trouble. Luckily, this insight is shared by the Chinese leadership. Increasingly, the rest of the world and China will be interdependent, while the power of China increases. The strive for a sustainable world will be adopted with ups and downs in most societies. The USA and the EU can catch up quickly once they understand the potential of Triple P growth. Therefore, Green China will meet countervailing power forcing the Chinese leaders to adopt sustainable world leadership. This is the consequence of climbing the new green growth curve.

Sustainable leadership for governing the new green growth curve requires numerous new solutions in government and private sector. The once successful Chinese decisive one-man show accompanied by selected firms needs to be replaced by organized thousands flower blossoming policies. Only then, the required scale and scope of solutions will become available.

The internationalization of the Chinese economy forces Chinese organizations to behave differently. The current common practice of people and network (guanxi) management in China will not be as effective in other markets. An increased focus on market demand and customers are necessary ingredients for success. Simultaneously, the principle of management by trust needs at least to be complemented by management by competence. This step is necessary not only because of foreign markets but also because of the phase of creating green innovative solutions.

Staying successful in the green business requires a combined vertical and horizontal innovation process. The five-year plans can still function as a strategic blueprint, but they need cross industry and bottom-up initiatives. Both strategic management and risk management of newly born ideas are needed instead of the current regular gambling behavior. The latter refers to the rush of Chinese businessmen not to miss markets and invest in markets as soon as first successes appear. It is typical for bonanza-like growth and perceived market opportunities, and the Chinese green sector is part of it.

The increased demand for capital to invest in social and ecological programs will force Green China to make adequate economic investments. As soon as investments in social security and ecology are needed in order to realize the desired harmonious society, less capital is available for reckless investment. The investments in the social and ecological area will show identical behavior of overinvestment and me-too investments. Bad loans will stay a big problem in the Chinese financial system. Therefore, better planning combined with risk management will play a bigger role in Chinese strategic innovation.

Chinese are masters in strategic thinking and planning. Their international expansion is based on grand strategic plans for acquisition of both government and business but lacks attention for operational and post-merger integration issues. In this respect, the grand plan-driven mergers and acquisition of Western activities were not always successfully managed. One of the main reasons is the lack of attention to post-merger integration issues leading to the desired bottom line results. Chinese pay attention to the big picture (the strategy or the acquisition) but do not always pay attention to the detail (daily operations). The value conversion of strategic innovation to buy Western brands and green technology often had a negative value because the integration of governance and leadership issues were largely ignored or treated wrongly.

Based on the issues in Fig. 8.3 and Table 8.3, the strategic innovation agenda for the People's Republic of China can be as follows:

1. The five-year plans need midterm reviews to be able to adapt to new circumstances.
2. The implementation of five-year plans in the provinces needs coordination in order to get complementary activities instead of copy–paste programs leading to overcapacity.
3. Create national innovation system based on strategic innovation and in particular nontechnical innovations instead of high tech.
4. Foster institutionalized innovation within individual private and public organizations.
5. Apply Triple P principle in global sourcing strategy.
6. Organize bottom-up innovation processes within organizations and apply the organized 1,000 blossoms strategy to unleash creative power for innovative solutions.
7. Synthesize the bottom-up processes with central planning.
8. Identify and manage cross-cultural issues in global expansion strategies.
9. Becoming number one superpower implies at least a number two or possibly a number three must be left and survive. Redefine Chinese win–win in a global win–win or a tri-star win–win.
10. Use the methodology of strategic innovation to create successful business outside China. Chinese business models need to be adapted for long-term global involvement and competition based on high-added value instead of low-cost high volume.

The biggest challenge for China is to continue their path of growth and development in a sustainable way. Copycat behavior and fast follower strategies can be organized by government. Strategic innovation cannot. New growth and continued innovation can only be achieved by human creativity and innovative entrepreneurialism. Bottom-up initiatives and decision processes are linked to the phase of indigenous innovation. Paradoxically, for China, further growth is dependent on the social aspect of sustainability. If China is not able to cope with this issue green will turn into black. History has shown these fallbacks are no exception in China, and it is the task of their leaders to prevent such a return to old times. Chinese leaders have to refocus from technology to soft skills and creating and allowing Chinese forms of bottom-up processes. The biggest leap forward they have ever made.

The items of Figs. 8.1, 8.2, and 8.3 are not new; they are elaborated in the previous chapters. But they force us to rethink about current positions and prepare for the future. For all parties involved, solving paradoxical situations characterize our current crossroad. The agenda issues for the strategic innovations are not only for government but business leaders as well. The core issue is that governmental and private business leaders should talk with each other on an equal level. Very often, in a European context, governmental leaders and business leaders communicate with each other from different institutions and expectations. For example, the government should make less regulations and implement more favorable conditions for private industries. In practice, policy makers have no idea what are really favorable conditions and what are not because most of them have no business experience. The other way round also holds. Both groups make their strategic plans independently of each other without knowing it from each other besides information given by industry lobby groups.

In the USA, the market-based innovation approach is the core of American capitalism and past successes. But American fossil oil industry has deep-rooted governmental relations which go beyond the free market mantra. The sustainable growth curve and Green China cannot be built on illusionary or dogmatic free market mechanisms. New governmental roles and involvement in green market capitalism is needed.

The case of renewables and electrical cars demonstrates the necessity to sit around the table and formulate a common strategy. Moreover, Green China requires to make radical shifts in corporate strategies and governmental policies. With regard to Green China, a few crucial additional elements have to be added. These additional elements boil down to paradoxes.

8.6 From Paradoxes to Solutions

Looking at Tables 8.1, 8.2 and 8.3, the directions of the strategic innovation for the USA, the EU, and China are actually paradoxes. We all have to be able to deal with paradoxes to renew. Finding the paradoxes and finding a way to use the paradoxes instead of being paralyzed by them will open up the black box of innovation.

With regard to strategic innovation in the context of Green China, many of the paradoxes have to be synthesized. Examples are symbiosis between private companies and government, free market versus central planning, government-owned companies versus privately held companies, and so on. It is impossible to identify them all, and it is even more impossible to give standard solutions because the synthesis depends on context and time. But paradox synthesis does offer the methodology to create solutions to countervail Green China.

China is a paradox. At least in the eyes of Westerners. The Western eyes are used to dichotomous world view like good and bad guy, binary 0-1 system, wrong versus right, and you are for or against us. Contrary to Chinese, the majority of the Westerners have difficulty to be flexible and to deal with both worlds at the same time. In all the previous chapters, the way Westerners and Chinese deal with greening the world has been explained along the lines of strategic innovation, business models, policies, and the different approaches. Apart from the content, the European response on Green China should also take into account the paradoxical interpretation of growth and development by the Chinese.

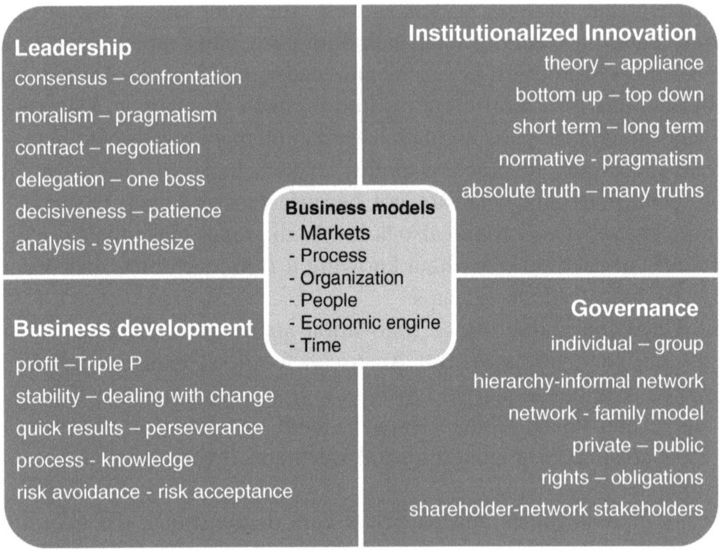

Fig. 8.4 Typical Green China paradoxes

The European Onion showed that in all the policies for unification, harmonization, regional development, and many others, the central part failed: strategy for growth. Based on recent European history, it is an illusion to think we could achieve a political union. But because of the rise of Green China, we need a coordinated countervailing power. This is the European paradox to be solved: standardization and complete unification against European diversity and differences.

The Americans fail on having a history of sustainability, the sense of urgency for sustainability, the insight that Triple P value creation replaces short-term profit making, the waning power of military force relative to global economic relations and access to resources, and the rise of economic power within the boundary of sustainable development. The American paradox to be solved is innovative capitalistic entrepreneurship against the next wave of prosperity based on sustainability.

The Chinese have to put strategic innovations and especially nontechnical innovations at the core of their policies and business strategies. The Chinese paradox is introducing decentralized innovation and bottom-up initiatives against central planning.

The European paradox has two dimensions. Without forceful governance, no political union or complete unified Europe is impossible. The other is that the extreme diversity and differences between European countries will not survive against Green China. The solution is to apply the method of paradox synthesis (van Someren 2005a).

China is greening because the Chinese are good at paradox synthesis. In Fig. 8.4, only a few typical Western–Chinese paradoxes which play important roles in generating strategic innovations toward a sustainable future are presented. The Chinese politic leadership found a way between communistic moralistic and capitalistic pragmatism to let people get rich. They turned the paradox of profit against planet into profit and planet by comprehensive green industry policies. The symbiosis between public and private sector creates big advantages for Chinese companies. Chinese companies use short-term tactic combined with long-term strategy to enter the Western green market. Because they can live with many truths, different green technologies (from the West) can be tried out at the same time to solve one problem. In the end, they may discover and create their own technology which is a combination and improvement of all the other technologies.

However, there is a long way to go for the Chinese to be really innovative, too. The Chinese model can also be a danger to a sustainable future of the world. For example, with regard to leadership, the Chinese leaders can be too decisive. The future of a nation or a company is too often too dependent on good leaders. But one should not forget that because of the increasing complexity of our society, not all knowledge can be united in one person. Therefore, the bottom-up process has the potential to be better because big mistakes can be avoided and creative solutions can be added. Many European leaders have to involve their subordinates before taking any decision. But the balance between bottom-up and top-down needs to be kept. The European leaders are too often not decisive enough. In time of crisis, they may be forced to take the right decision among all the different suggestions from bottom-up. There is still chance for a good leader if one sees the paradox and can deal with it.

Another example is the paradox between change and steadiness. Chinese are used to deal with change. That is the reason that they do not always commit to one solution but keep open alternatives. It is also the reason that they sometimes waste resources on meaningless or risky environmental projects. On the contrary, many Westerners strive for risk avoidance, stability, and security. That is why the quality and efficiency of environmental projects are much higher. Once the West combines the steadiness with taking strategic market chances, Green China will be the growth wave which the West can ride too. The paradox of change and steadiness shows also in dealing with contracts. The West is good at written contracts. This is also a form of making explicit the rights, but apart from rights, one has obligations. The Chinese only want to trust their family and friends with business because they rely on obligations to each other. This limits the business in many ways. Playing the card of obligations together with rights can be very fruitful when dealing with Chinese. The discussion about IP-rights in China shows this very clear. Bringing forward the IP-issue based on rights does not help unless the responsible (local) government realizes their obligations. But as soon as the Chinese are involved in the mutual development of IP-rights and they have to be applied and sold outside China, the picture can change completely. Their success will also base on their sense of both rights and obligations. Because they cannot play anymore on their home turf and will be more dependent on the Western business partner, they will have personal interests and obligations accordingly.

In the case of business model, the very common Chinese family model versus the Western network model shows a big paradox. The differences form the main market vehicles for doing business. Many differences between Western and Chinese business model are paradoxes. In Table 8.4, a few paradox features are described.

Table 8.4 Western network model versus Chinese family model

Western network model	Family model
No clear leader coordinator instead	Leadership based on rank or age
Open communication	Communication with trustee
Process oriented	Whole-picture orientated
Individual initiation important	In-group harmony important
One for oneself	"One for all" and "all for one"
Task orientated	Team result orientated

There is a market for the green sector in China, but the West is not earning money from it. The wrong business model is the most common cause of failures of the Western business in China. When Western and Chinese organizations cooperate or form an alliance, very often the general notions from Table 8.4 are helpful to describe the alliance behavior. In the Chinese family-based model, leadership is based on rank or age contrary to the Western situation based on task for coordination. When the role of the Western leader is not clear, market chances will be missed when the Chinese do not listen to the boss. On the other hand, headquarters have difficulty to be the boss without risking the Chinese to lose face. In this respect, also the open communication between employees with the right to speak based on their rank or valuable creative contributions contrasts with the Chinese communication mainly based on a trustee. Western managers govern their alliance based on a process to get things done or delivered according to planning. For the Chinese counterpart, the whole picture is important. The Western management should find the balance between forcing the Chinese counterpart to comply with the planning and flexibility. The other aspects relate to the individual versus collective way of thinking. Traditionally, group harmony is more important than individual contributions. But in relation to innovation, this might be one of the barriers. Western companies which move their research center to China may be disappointed. The innovation capacity generated by individual freedom is the bottleneck for the green sector. The West still has a chance to win the innovation battle.

To find a way in the greening China, paradoxes should be noticed, analyzed, and responded. The description and appliance of these Western–Chinese paradoxes should be treated with care because depending on the situation, a complete reversed interpretation is possible. For instance, when a great opportunity comes by, Chinese businessmen can react extremely fast and decide within seconds to buy a rival when in trouble. This kind of reckless behavior also caused the gamble overinvestment of the Chinese green sector. This is the chance for the much more mature Western way of doing business. When it is combined with the Chinese model, it can be undefeatable.

When we take together the issues of threats and opportunities, mutual dependency, and paradoxes, then we have to conclude that in each individual situation, the strategic innovations have to be created, developed, and tailor-made. This holds for both private and public organizations or their symbiotic working together.

8.7 Continuous Innovation with Strategic Innovation Cycle©

The multi-star constellation, as described in the previous section, forces us to reinvent our self much more often than we were used to in the Western context. Chinese conditions change rapidly and frequently due to adapted or new governmental rules, new customer needs and requirements, new rivals, and new megatrends to reckon with. Only with the help of strategic innovations we can safely continue our journey through the economic interstellar cosmos of Green China, other BRIC countries, and the USA.

Most Western China experts quote each other in doing business with China. One-sided solutions were presented. Too much attention was paid to Chinese etiquettes instead of to what really matters. Strategic insights and answers are things we should pay attention to before we deal with China. The right strategic answer depends on the situation. The same holds for a strategic answer on Green China and the Green Dragons. We need to be innovative all the time to have strategic answers depending on the situation. For Western readers, this is a very disappointing answer. But as soon as you understand Green China, you value this way of thinking. Westerners want clear and unequivocal recommendations. Sometimes they might work, but in most instances not. In fact, every time we need unique answers, there is no recipe. From the perspective of competition, one should be excited now. Because only those who accept the necessity to find new approaches to do business will survive and earn money.

The Strategic Innovation Cycle© offers an approach to adapt to new situations and create the necessary strategic innovations (see Fig. 8.5).

The triple star constellation demonstrated a fast ever-changing unpredictable situation. For business managers and governments, it is necessary to renew your corporate strategies and policies. The Strategic Innovation Cycle© enables you to create adequate answers on the rise of Green China.

In the tri-star constellation, the EU, the USA, and China have to apply the Strategic Innovation Cycle© in order to prepare themselves for the road along the new long growth cycle of sustainable development. In the next decades, it will be decided who will win the race. Will history repeat itself and new dominant countries with new dominant firms will foster new technologies, organizational forms, and new governance structures? Or will a new history be developed which deviates from the past?

8.8 Strategic Innovations Will Create the Next Superpowers

Green China is part of a once in a lifetime strategic innovation toward a new industrial long growth cycle for the rest of this century. The previous long cycles were characterized by the domination of one particular country such as the USA in the past decades. Past cycles were also accompanied by new technologies, new organizational forms, new political structures, new institutions, and new dominant firms. When you believe in repetition and extrapolation, China will indeed be the

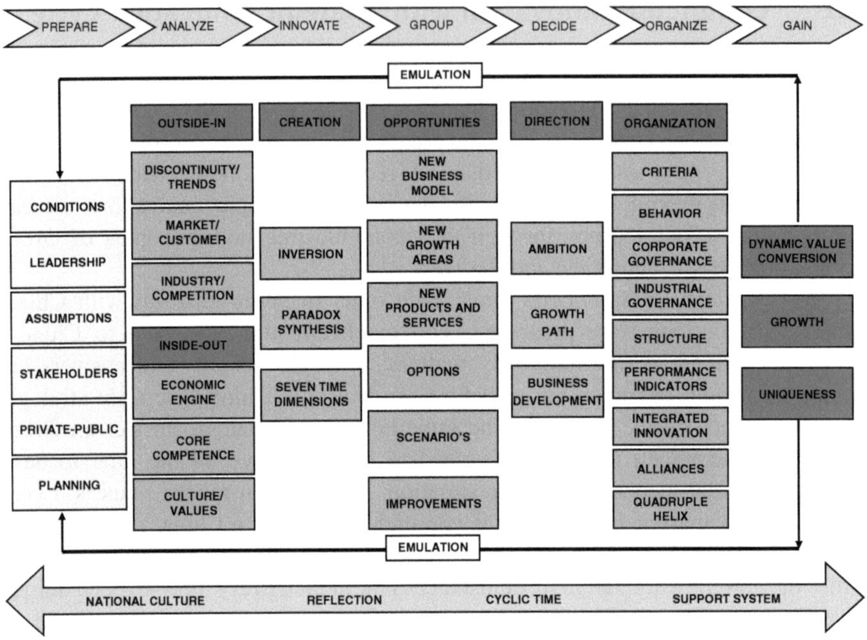

Fig. 8.5 Strategic innovation cycle©. (Source: van Someren, 2005)

hottest candidate to become the new dominant power. But what we tried to demonstrate is that China always has at least two faces. The development and growth of China is both a threat and an opportunity for others. But also for China itself. China needs many strategic innovations in government and corporations. Some of these strategic innovations and in particular, the nontechnical innovations such as bottom-up processes amplified by social media will create a turmoil in all groups of society. Moreover, sustainability issues are great challenges for China. The USA, the EU, and other regions still have the opportunity to catch up as equal rivals of China. The good news is that the Western private sector has an innovation tradition. The bad news is that the public sector does not play the right role yet. For a sustainable future, strategic innovations need to be created both in private and public organizations. The USA and the EU need to create strategic innovations that take into account all the paradoxes and their specific Chinese ways of doing business. The Chinese need to do the same when entering other markets. Only then will dealing with Green China be successful. Therefore, both the EU and the USA have great opportunities to continue their success and green Europe and green USA can become new reborn superpowers as well. The absolute worldwide key challenge is the creation and implementation of strategic innovations. Strategic innovations decide about greening of industry and about the next superpower; it may be Green China.

References

Someren, T. C. R. van & van Someren-Wang, S. (2011). Kans na harde landing. In: *Financieele Dagblad* (p. 7). 2 Dec 2011.

Someren, T. C. R. van (2005a). *Strategische Innovationen. So machen Sie Ihr Unternehmen einzigartig*. Wiesbaden, Germany: Gabler Verlag.

MIX
Papier aus verantwortungsvollen Quellen
Paper from responsible sources
FSC® C105338

If you have any concerns about our products,
you can contact us on
ProductSafety@springernature.com

In case Publisher is established outside the EU,
the EU authorized representative is:
Springer Nature Customer Service Center GmbH
Europaplatz 3, 69115 Heidelberg, Germany

Printed by Libri Plureos GmbH
in Hamburg, Germany